AF722953

Taxonomy of *Polygonum* Section *Echinocaulon* (Polygonaceae)

Chong-Wook Park

QK Memoirs of the New York
1 Botanical Garden ; v.
.N525
v. ct/t

The New York Botanical Garden
Bronx, New York 10458, U.S.A.
Issued: 20 July 1988

MEMOIRS OF THE NEW YORK BOTANICAL GARDEN VOLUME 47

QK
1
.N525
v.47

Copyright © 1988

The New York Botanical Garden

Published by
The New York Botanical Garden
Bronx, New York 10458

International Standard Serial Number 0071-5794

Library of Congress Cataloging-in-Publication Data

Park, Chong-Wook, 1953-
Taxonomy of Polygonum section Echinocaulon (Polygonaceae) / Chong-Wook Park.
p. cm.—(Memoirs of the New York Botanical Garden; v. 47)
Bibliography: p.
Includes index.
ISBN 0-89327-329-5
1. Polygonum—Classification. 2. Botany—Classification. I. Title. II. Title: Echinocaulon (Polygonaceae) III. Series.
QK1.N525 vol. 47
[QK495.P78]
581 s—dc19
[583'.917] 88-5361
CIP

Copyright © 1988 The New York Botanical Garden

International Standard Book Number 0-89327-329-5

JULY 1988 MEMOIRS OF THE NEW YORK BOTANICAL GARDEN **47:** 1–82

Taxonomy of *Polygonum* Section *Echinocaulon* (Polygonaceae)

Chong-Wook Park[1]

Contents

Summary

Park, Chong-Wook (New York Botanical Garden, Bronx, New York 10458, U.S.A.). Taxonomy of *Polygonum* section *Echinocaulon* (Polygonaceae). Mem. New York Bot. Gard. **47:** 1–82. 1988. *Polygonum* section *Echinocaulon,* as delimited here, includes 21 species. The section is usually characterized by the occurrence of recurved prickles on the angles of the stem, petioles, and abaxial surfaces of major leaf veins. The species of section *Echino-*

[1] Present address: Harvard University Herbaria, Harvard University, 22 Divinity Avenue, Cambridge, Massachusetts 02138.

caulon are found in Asia, eastern North America, South America, eastern Australia, and southeastern Africa, with the center of diversity in eastern Asia. Section *Echinocaulon* shows complex disjunct distribution patterns; there are strong intercontinental relationships in the distribution of the section between eastern Asia and eastern North America, and between South America and southeastern Africa. A taxonomic treatment based on comparative morphology, biogeography, cladistic analyses, and flavonoid chemistry is provided. Descriptions of all the taxa and a key to the species and varieties are presented. The descriptions presented here are based on field observations and the examination of some 4000 specimens from the major herbaria of the world. The treatment includes one new combination: *P. dichotomum* var. *angustissimum* (Hook. f.) Park.

Introduction

Polygonum is a taxonomically difficult, highly variable genus of about 300 species. Plants of *Polygonum* are annual or perennial herbs, shrubs or vines of moist habitats, many becoming tenacious weeds in disturbed areas. The genus includes several natural groups, the taxonomic ranks of which are much in dispute. Meisner (1826, 1832, 1855, 1856, 1865) and many other authors (Graham & Wood, 1965; Small, 1892, 1895; Steward, 1930) treated these groups as subgenera or sections of *Polygonum* s. lat., whereas a number of others (Gross, 1913; Haraldson, 1978; Löve & Löve, 1956; Nakai, 1926; Rafinesque, 1837) regarded at least some of them as distinct genera. However, the overlapping of major characters makes subgeneric classification difficult and complicated.

Section *Echinocaulon,* first recognized by Meisner (1832), consists of 21 species that usually occupy moist, marshy, or swampy habitats. The section is usually defined by the occurrence of recurved prickles on stems, petioles, and abaxial surfaces of major leaf veins. Additional characters separating this from other sections of *Polygonum* include a sprawling to climbing habit, the presence of stellate and tufted hairs (Haraldson, 1978), and elliptic to triangular leaf shapes, frequently with sagittate to hastate bases. The species of section *Echinocaulon* are found in Asia, eastern North America, South America, eastern Australia, and southeastern Africa, with the greatest concentration of species in eastern Asia.

Several authors (Danser, 1926; Meisner, 1826, 1832, 1855, 1856, 1865; Small, 1895; Steward, 1930) provided morphological descriptions and keys to some species of section *Echinocaulon* in monographs of *Polygonum* or Polygonaceae. Otherwise the section as a whole has received little taxonomic attention.

As in other sections of *Polygonum,* most of the species in section *Echinocaulon* show very complicated patterns of morphological polymorphism, resulting in taxonomic confusion and difficulty in delimiting species boundaries and determining relationships. These problems suggested the need for a modern taxonomic revision with a focus on the degree of morphological diversity and phylogenetic relationships. The main objective of this study is to provide a taxonomic monograph of *Polygonum* section *Echinocaulon,* based on comparative morphology, biogeography, cladistic analyses, and comparative flavonoid chemistry. Ancillary studies associated with the present investigation resulted in publications on nomenclatural changes (Park, 1986a), lectotypifications (Park, 1986b), and comparative flavonoid chemistry (Park, 1987).

Morphology

Descriptions of the morphological structures in the following pages are based on field observations, and on ca. 4000 specimens from the major herbaria of the world and those collected by the author in Korea and the United States.

The species of section *Echinocaulon* are ascending to scandent, annual herbs arising from fibrous root systems. Stems, petioles, and abaxial surfaces of major leaf veins are retrorsely prickly, a characteristic by which the section is usually defined. Plant height usually ranges from 0.3 to 1.5 m at maturity. The largest species, particularly *P. perfoliatum* and *P. senticosum,* are scandent, frequently 4–6 m tall, and climbing over other objects by means of recurved prickles.

Stems

Stems of section *Echinocaulon* are branched and armed with recurved prickles along the angles. As in many other *Polygonum* species, nodes are conspicuously swollen and lower ones frequently produce fibrous adventitious roots. In *P. thunbergii,* lower nodes often produce flowering and fruiting stolons. The stem habit is variable, from decumbent to ascending or scandent.

Leaves

Leaves of section *Echinocaulon* are alternate, entire, and simple with ciliate to prickly margins; abaxial surfaces of major veins are usually prickly. Leaf shapes encountered in the section are illustrated in Figures 1 and 2. The leaf length/width ratio and the petiole length/leaf length ratio for each taxon are provided in Table I. Leaf shapes are diverse, ranging from linear lanceolate to broadly triangular. The leaf shape usually shows considerable intraspecific variation, but it is still valuable in recognizing some species. In *P. perfoliatum, P. senticosum,* and *P. debile* (Fig. 1: a–d), leaves are broadly triangular with truncate to weakly hastate bases, an unusual leaf shape in the section. In *P. arifolium, P. thunbergii,* and *P. biconvexum,* leaves are, in general, triangular in outline, but they tend to have deeply hastate bases with divergent, triangular basal lobes (Fig. 1: e–h). In addition, the middle lobe tends to be broadly elliptic rather than triangular. In these species, leaves are usually long-petioled, and petiole length/leaf length ratio usually ranges from 0.45 to 1.17 (Table I). In the rest of the species, the leaf shape becomes far more diverse, ranging from linear lanceolate to ovate with weakly cordate to sagittate bases (Fig. 2). Intraspecific variation in leaf shape is also more extensive in these species. The petiole length is variable, but seldom exceeds the petiole length/leaf length ratio of 0.40 (Table I).

Ocreae

Ocrea morphology provides some of the most useful characters in *Polygonum* taxonomy (Meisner, 1826, 1856; Steward, 1930). Ocrea shapes found in the section are illustrated in Figure 3. In section *Echinocaulon,* shape, texture, pubescence, and presence or absence of cilia at the tip are especially useful. Ocreae of *P. perfoliatum, P. senticosum, P. debile, P. thunbergii,* and *P. biconvexum* are either foliaceous or have foliaceous limbs (Fig. 3: 1–7). In the remaining species, ocreae are completely scarious, and funnelform to cylindrical; tips are either truncate or oblique (Fig. 3: 8–25). The pubescence of ocreae shows intraspecific and age-related variation, but still is useful in recognizing species. The pubescence of ocreae in the section is summarized in Table II.

Inflorescences

Some inflorescence types encountered in section *Echinocaulon* are illustrated in Figure 4. Inflorescences arise from the axils of upper leaves and directly from stem apices. They consist of a series of fascicles of minute, short-pedicellate flowers disposed in a somewhat spikelike or headlike cluster, terminating an elongated inflorescence axis. Each flower fascicle is usually two- to four-flowered and subtended by a small bract. Each flower also bears paired bracteoles that are usually broadly elliptic and fused at the base (Fig. 4: A, D). In most species, inflorescences are elongate and branched several times. One or two lateral inflorescence branches arise from the nodes on the main inflorescence axis, and repetition of this branching pattern results in panicles of three to eight clusters of flower fascicles. Each inflorescence is subtended by an inflorescence bract that resembles the ocrea. In *P. perfoliatum,* inflorescences are unbranched, and the flower fascicles are spicate along the relatively short inflorescence axes (Fig. 4: C). In this species, one to two flower fascicles also are borne in the axils of upper reduced leaves, but they generally fail to develop elongated inflorescence branches.

There is considerable variation in the arrangement of flower fascicles. In most species, flower fascicles are closely spaced and form a dense spikelike or headlike cluster (Fig. 4: A). In a few species, as in *P. praetermissum, P. dissitiflorum, P. subsagittatum,* and *P. rubricaule,* they are widely spaced, forming a loose, interrupted panicle. In addition, the bracts that subtend flower fascicles are somewhat funnelform, and completely surround the inflorescence branch (Fig. 4: D, E).

The sequence of flowering of an inflorescence roughly proceeds from base to apex. The flowers

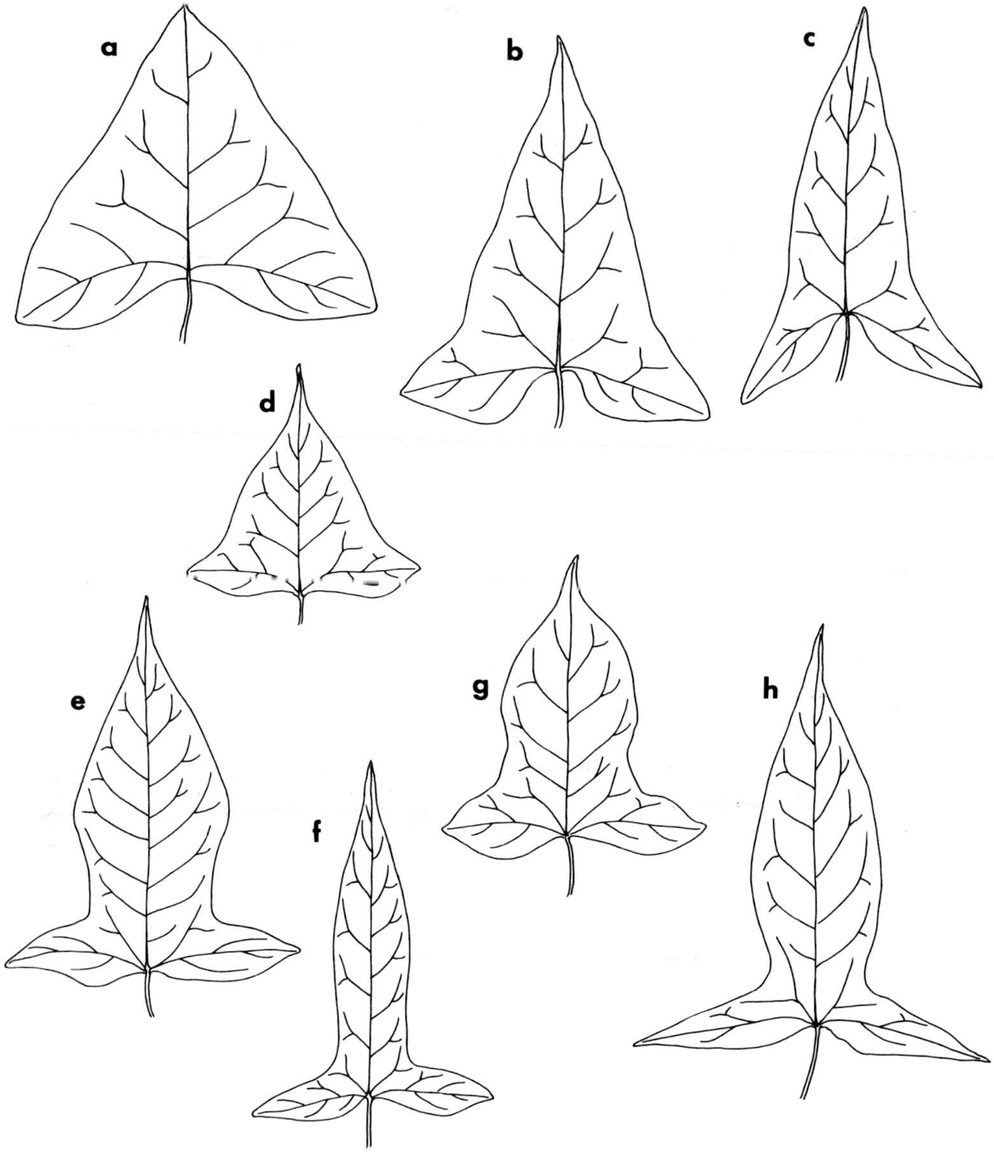

Fig. 1. Representative leaves of *Polygonum* section *Echinocaulon*. a. *P. perfoliatum*; b. *P. senticosum* var. *senticosum*; c. *P. senticosum* var. *sagittifolium*; d. *P. debile*; e. *P. thunbergii* var. *thunbergii*; f. *P. thunbergii* var. *maackianum*; g. *P. biconvexum*; h. *P. arifolium*. Scale 1:1.4.

within a fascicle mature in succession, and often the first one has completely matured before the next one appears above the bract margin.

FLOWERS

Flowers of the section *Echinocaulon* are bisexual and regular, with a slightly variable number of parts. Important species characters include the number of tepals and styles, and the number and arrangement of stamens. The perianth is petaloid, whitish-green to reddish-purple, and persistent; it encloses the achene at maturity. In *P. perfoliatum,* the perianth becomes fleshy and blue in fruit, an unusual trait in the genus. As in other sections of *Polygonum,* the perianth is four- or

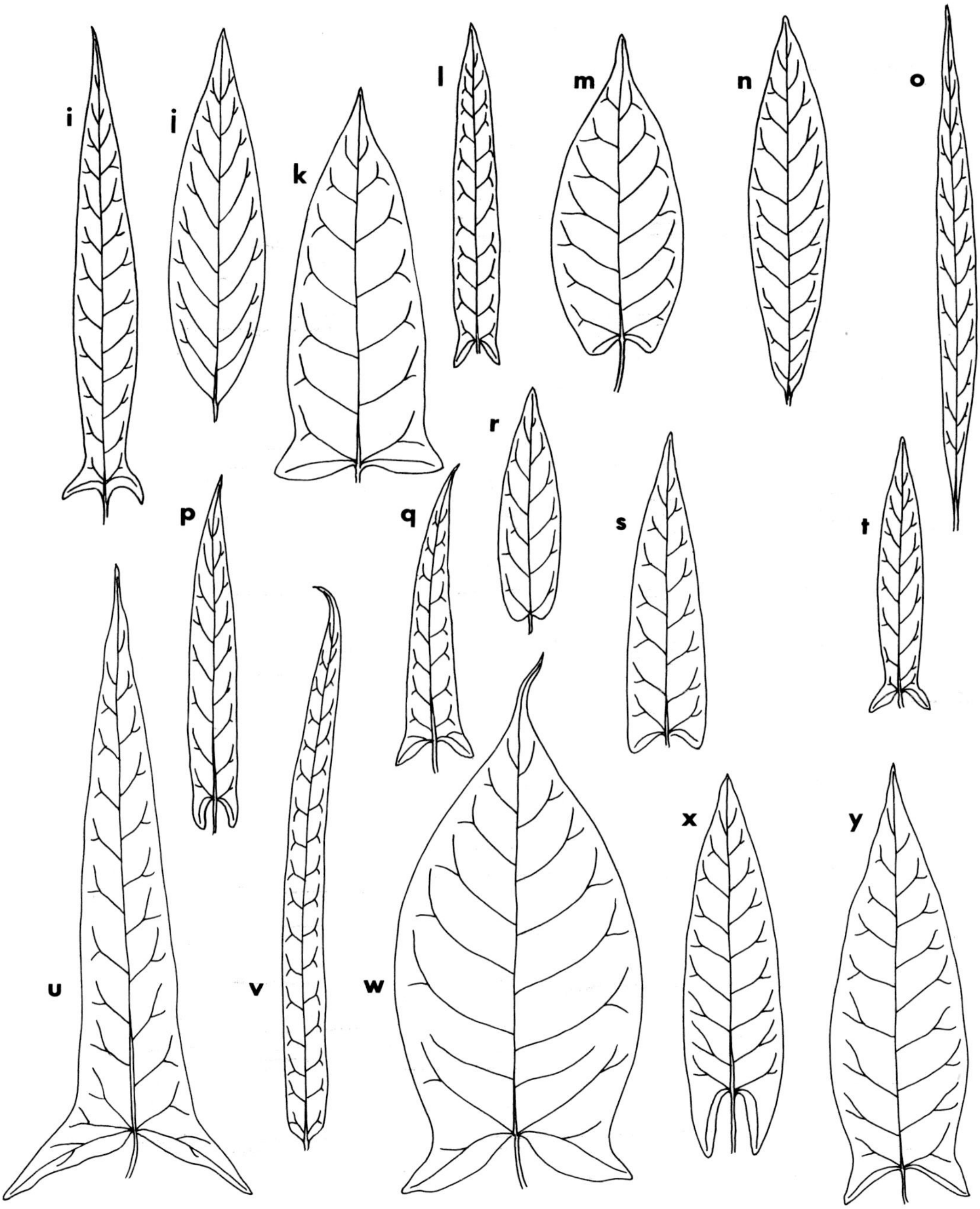

Fig. 2. Representative leaves of *Polygonum* section *Echinocaulon*. i. *P. brachypodum*; j. *P. bungeanum*; k. *P. strigosum*; l. *P. clarkei*; m. *P. muricatum*; n. *P. dichotomum* var. *dichotomum*; o. *P. dichotomum* var. *angustissimum*; p. *P. rubricaule*; q. *P. subsagittatum*; r. *P. breviochreatum*; s. *P. meisnerianum* var. *meisnerianum*; t. *P. praetermissum*; u. *P. stelligerum*; v. *P. meisnerianum* var. *beyrichianum*; w. *P. dissitiflorum*; x. *P. sagittatum*; y. *P. hastatosagittatum*. Scale 1:1.5.

Table I

Leaf length/width ratios and petiole length/leaf length ratios of *Polygonum* section *Echinocaulon* species. Numbers given are mean ± standard deviation. Measurements were taken from herbarium specimens (n = 10–20)

Taxa	Leaf length/width	Petiole length/leaf length
P. perfoliatum	0.88 ± 0.14	1.17 ± 0.17
P. senticosum		
var. *senticosum*	0.95 ± 0.06	0.83 ± 0.11
var. *sagittifolium*	2.20 ± 0.29	0.72 ± 0.15
P. debile	0.89 ± 0.11	0.57 ± 0.09
P. thunbergii		
var. *thunbergii*	1.40 ± 0.12	0.45 ± 0.10
var. *maackianum*	1.29 ± 0.31	0.66 ± 0.19
P. biconvexum	1.13 ± 0.13	0.57 ± 0.12
P. arifolium	1.23 ± 0.15	0.52 ± 0.13
P. stelligerum	4.12 ± 0.51	0.31 ± 0.07
P. muricatum	1.99 ± 0.29	0.24 ± 0.05
P. hastatosagittatum	3.57 ± 0.65	0.27 ± 0.06
P. strigosum[a]	3.19 ± 0.54	0.25 ± 0.08
P. breviochreatum	3.43 ± 0.55	0.10 ± 0.02
P. dissitiflorum	1.99 ± 0.17	0.35 ± 0.66
P. praetermissum	6.97 ± 1.67	0.08 ± 0.03
P. subsagittatum	6.21 ± 0.89	0.11 ± 0.03
P. bungeanum	3.91 ± 0.67	0.12 ± 0.03
P. rubricaule	6.67 ± 1.86	0.11 ± 0.03
P. sagittatum	3.77 ± 0.44	0.26 ± 0.09
P. meisnerianum		
var. *meisnerianum*	4.88 ± 1.25	0.06 ± 0.02
var. *beyrichianum*	9.72 ± 1.88	subsessile
P. brachypodum	5.90 ± 1.17	0.14 ± 0.05
P. dichotomum		
var. *dichotomum*	4.12 ± 1.20	0.11 ± 0.17
var. *angustissimum*	9.61 ± 2.05	subsessile
P. clarkei	5.89 ± 0.91	0.08 ± 0.02

[a] Australian population.

five-parted. The pentamerous condition is dominant in the section, tetramerous perianths being found only in *P. arifolium, P. stelligerum, P. praetermissum,* and *P. biconvexum.* According to Laubengayer (1937), a trimerous whorled condition is the fundamental plan in the Polygonaceae, and the tetra- and penta-merous conditions in *Polygonum* are derived from it by fusion or suppression in the development of tepals. The stamen number varies from four to eight, and shows considerable intraspecific variation in some species. Stamens are inserted at the base of the perianth in one or two whorls, alternating with small, flattened glandular teeth. The number of stamen whorls appears to be constant in each species. The gynoecium is unilocular, syncarpous, and contains a single basal orthotropous ovule. It is two- or three-angled, depending on species, and develops into a biconvex or trigonous achene, respectively. The former has a two-cleft style and the latter has a three-cleft one. Stigmas are minutely capitate.

Achenes

Achenes are trigonous or biconvex in cross section, and this character is relatively constant within each species. However, in *P. strigosum,* both types rarely occur on the same plant. The achene surface varies from smooth and shiny to rugose and dull, and shows intraspecific variation in some species. The color of the achene is usually dark brown to black, but in *P. thunbergii* var. *thunbergii,* achenes are much lighter, usually yellowish-brown.

Trichomes and Prickles

Trichomes of *Polygonum* have been examined by previous authors (Gross, 1913; Haraldson, 1978; Mitchell, 1971). Hairs encountered in section *Echinocaulon* are illustrated in Figure 5. Five types of hairs occur in the section: (a) simple unicellular hairs, (b) simple multiseriate hairs, (c) tufted hairs composed of four to seven short, ascending rays, (d) sessile stellate hairs with four to seven slender rays, and (e) multicellular glandular hairs with relatively long stalks. Table II summarizes the taxonomic distribution of hair types in the section. All the taxa except for *P. perfoliatum, P. rubricaule,* and *P. senticosum* var. *sagittifolium* develop at least one type of hair. The distribution of hair types shows considerable variation within species, and descriptions outlining the extent of intraspecific variation are provided in the taxonomic treatment.

Recurved prickles are borne along the angles of the stem, petioles, and abaxially on major leaf veins. Prickles are composed of elongate, lignified cells. In most species, prickles are relatively short (1.0–1.5 mm long), stout, and broad-based. In a few species, they are much longer (2–2.5 mm long), more slender, and bristly. The distinction between these two extreme types is not sharp, and intermediate forms are observed in some species. Prickle characters are considered in detail in the taxonomic treatment.

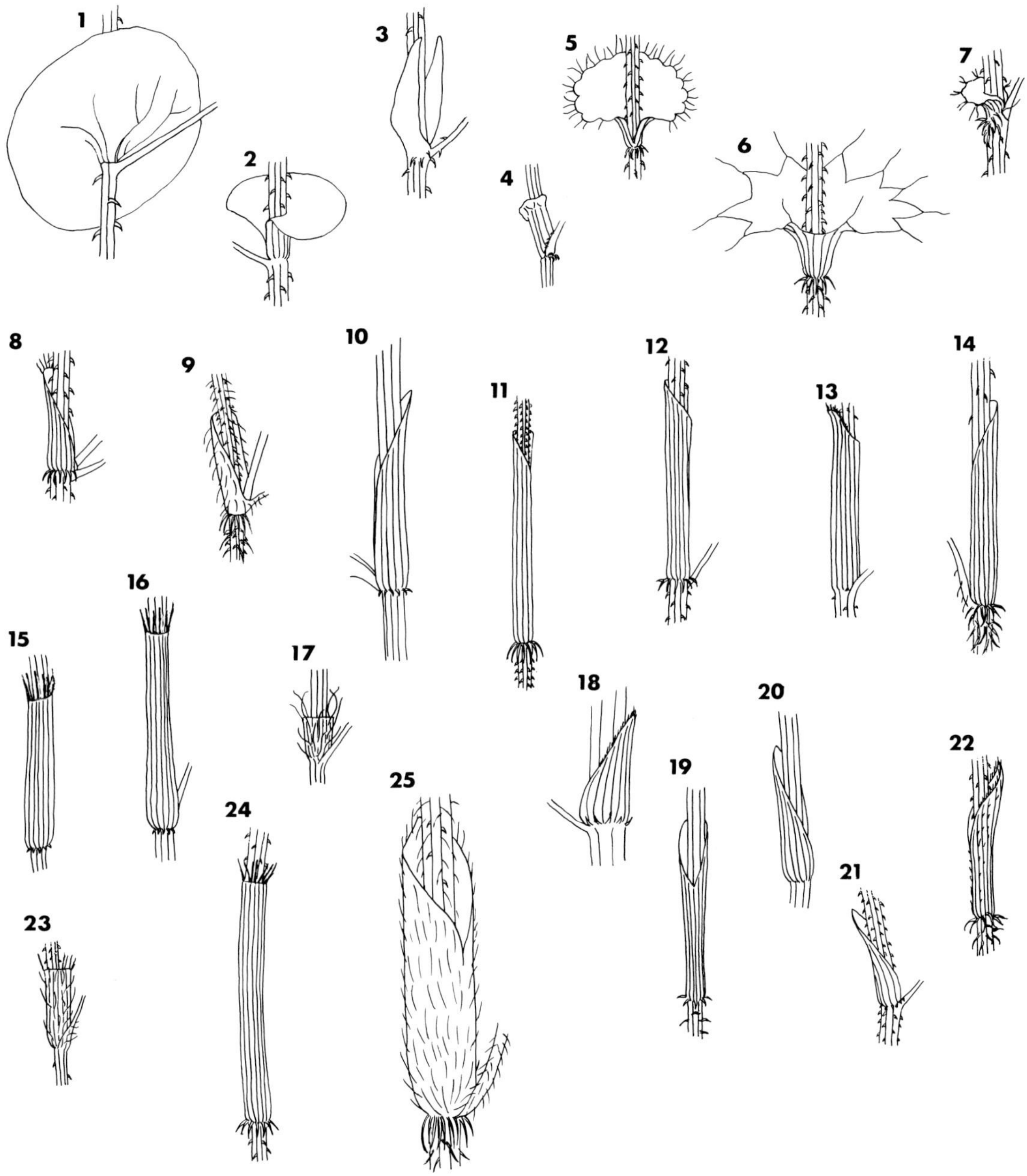

Fig. 3. Representative ocreae of *Polygonum* section *Echinocaulon.* 1. *P. perfoliatum*; 2. *P. senticosum* var. *senticosum*; 3. *P. senticosum* var. *sagittifolium*; 4. *P. debile*; 5. *P. thunbergii* var. *thunbergii*; 6. *P. thunbergii* var. *maackianum*; 7. *P. biconvexum*; 8. *P. arifolium*; 9. *P. meisnerianum* var. *meisnerianum*; 10. *P. meisnerianum* var. *beyrichianum*; 11. *P. clarkei*; 12. *P. dichotomum* var. *dichotomum*; 13. *P. dichotomum* var. *angustissimum*; 14. *P. brachypodum*; 15. *P. muricatum*; 16. *P. hastatosagittatum*; 17. *P. breviochreatum*; 18. *P. dissitiflorum*; 19. *P. praetermissum*; 20. *P. rubricaule*; 21. *P. sagittatum*; 22. *P. subsagittatum*; 23. *P. bungeanum*; 24. *P. strigosum*; 25. *P. stelligerum*.

Table II

Taxonomic distribution of trichome types in *Polygonum* section *Echinocaulon*. a: simple unicellular hair; b: simple multiseriate hair; c: tufted hair; d: sessile stellate hair; e: glandular hair

Taxa	Stem	Leaf	Ocrea	Inflorescence branch
P. perfoliatum	—	—	—	—
P. senticosum				
var. *senticosum*	a	a	—	e
var. *sagittifolium*	—	—	—	—
P. debile	—	b, d	—	e
P. thunbergii				
var. *thunbergii*	(d)	b, (d)	(d)	(d), e
var. *maackianum*	d	b, d	d	d, e
P. biconvexum	(d)	b, d	d	e
P. arifolium	—	b, d	(b)	d, e
P. stelligerum	b, d	b, d	b, d	b, d, e
P. muricatum	—	b, (c), (d)	—	a, e
P. hastatosagittatum	—	b, (c), (d)	—	(c), e
P. strigosum	—	b, (d)	(b)	(c), e
P. breviochreatum	—	b, (d)	b	b, e
P. dissitiflorum	d	b, d	—	b, d, e
P. praetermissum	—	—	(b)	(e)
P. subsagittatum	c, (e)	b, c	b	e
P. bungeanum	(e)	b	b	e
P. rubricaule	—	—	—	—
P. sagittatum	—	(b)	—	—
P. meisnerianum				
var. *meisnerianum*	c, (e)	b, (e)	b	e
var. *beyrichianum*	(c)	(b), (e)	(b), (e)	e
P. brachypodum	d	b, d	(b), (d)	b, e
P. dichotomum				
var. *dichotomum*	—	—	(b)	e
var. *angustissimum*	—	—	—	e
P. clarkei	—	b	b	e

Flavonoid Chemistry

Flavonoid chemistry of 13 species in section *Echinocaulon* was studied (Park, 1987). Thirty-six flavonoid compounds were isolated and identified, most of which were glycosylated and/or methoxylated derivatives of the flavonols kaempferol, quercetin, and myricetin, and of the flavones apigenin and luteolin. Among them, quercetin 3-O-glycosides were major flavonoid constituents, and present in all species. Most species of section *Echinocaulon* were readily distinguished by their flavonoid profiles. The flavonoid cladogram, based on nine flavonoid characters, suggested four basic chemical lineages in the section: (1) *P. perfoliatum* and *P. senticosum,* which are characterized by complete absence of flavones, (2) *P. sagittatum,* a chemically very distinctive species containing tricin together with both flavonols, kaempferol and quercetin, (3) *P. muricatum, P. hastatosagittatum, P. meisnerianum, P. praetermissum,* and *P. dichotomum,* which are characterized by predominance of flavonol O-glycosides over flavones, and (4) *P. thunbergii, P. strigosum, P. stelligerum, P. dissitiflorum,* and *P. arifolium,* a group of chemically very diverse species characterized by its complex arrays of C-glycosyl flavones and by frequent O-methylation. These findings will not be described in detail here, but references will be made to them where appropriate.

Biogeography

The species of section *Echinocaulon* are found in Asia, eastern North America, South America, eastern Australia, and southeastern Africa (Fig. 6). Most species are confined to temperate regions, but a few species occur in subtropical to tropical parts of these continents. The species of

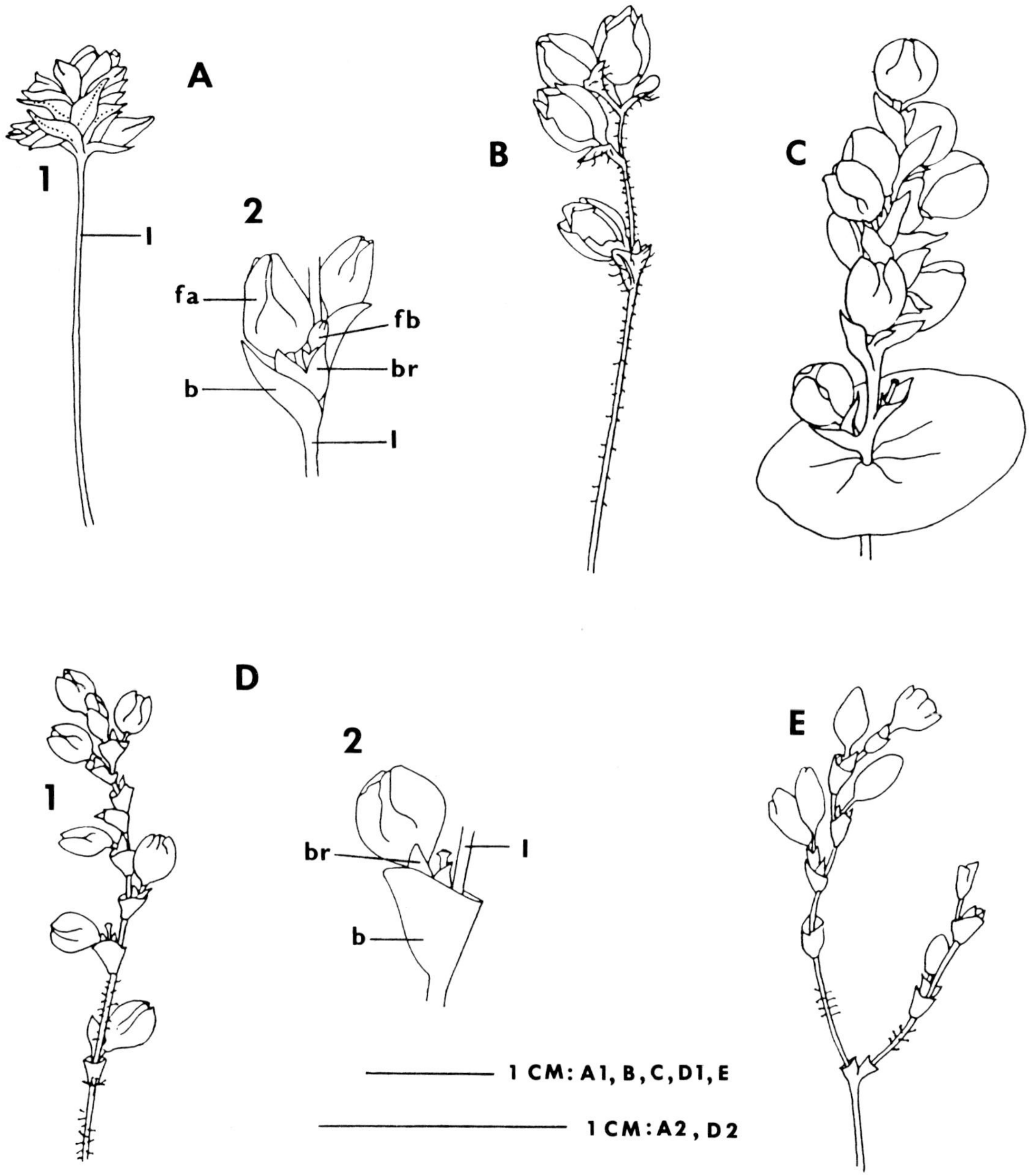

Fig. 4. Some inflorescence types encountered in *Polygonum* section *Echinocaulon.* A (1, 2). *P. sagittatum*; B. *P. arifolium*; C. *P. perfoliatum*; D (1, 2). *P. bungeanum*; E. *P. praetermissum.* b = bract; br = bracteole; fa = first flower; fb = second flower bud; i = inflorescence branch.

the section usually occupy moist, marshy, or swampy habitats, and often occur as weeds in disturbed areas. The section shows complex disjunct distribution patterns, and common taxa often occur in widely separated regions.

Species diversity maps of section *Echinocaulon* are shown in Figures 7, 8, and 9. The section is centered in eastern Asia, with the greatest concentration of species in eastern China, Korea, and southern Japan (Fig. 7). The ranges of 14 of 21 species in the section converge in this area, suggesting that eastern Asia is the principal re-

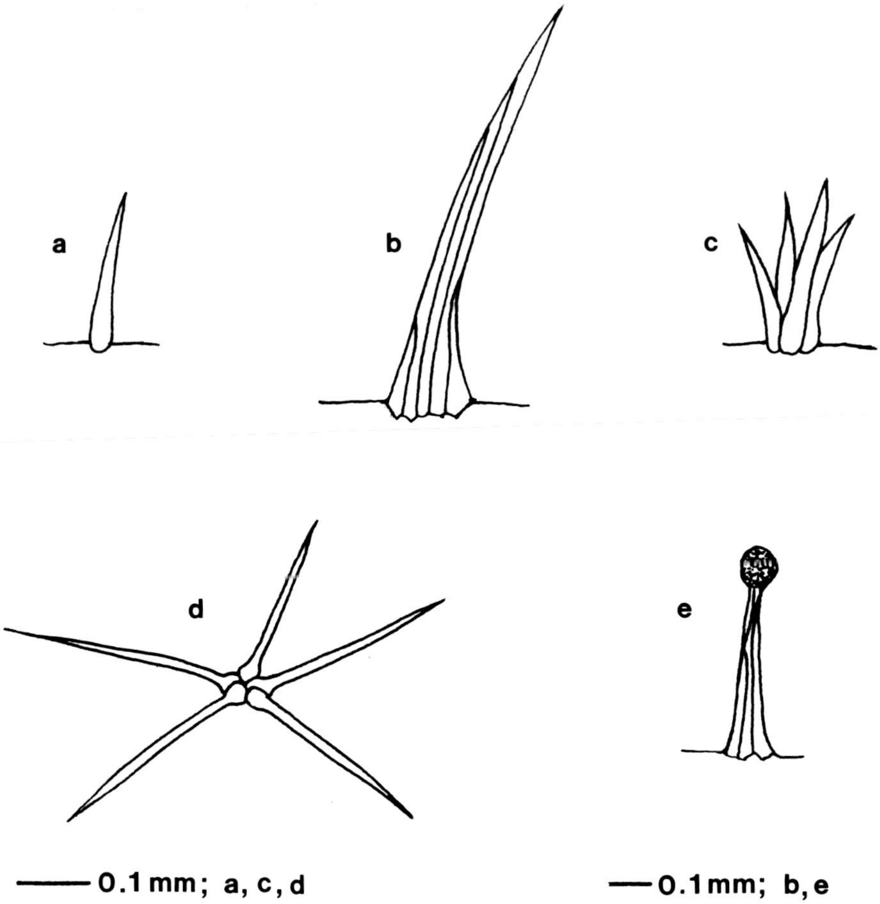

Fig. 5. Trichome types found in *Polygonum* section *Echinocaulon.* a. simple unicellular hair; b. simple multiseriate hair; c. tufted hair; d. sessile stellate hair; e. glandular hair.

gion of diversification of the group. Among the species found there, four have distributions that extend well beyond the region; *P. perfoliatum* in southeast Asia, *P. sagittatum* in eastern North America, and *P. dichotomum* and *P. praetermissum* in eastern Australia.

The complement of species found in Australia includes one additional taxon, *P. strigosum.* Like *P. dichotomum* and *P. praetermissum* encountered in Australia, *P. strigosum* also is widely distributed in Asia: from Nepal and northern India through southern China and southeast Asia to eastern Australia, including Tasmania. The continuous distribution ranges of these species between Asia and Australia and the lack of endemic species in Australia suggest that they may have migrated from continental Asia through the intervening islands of southeast Asia, or across the Sunda Shelf during periods of low sea level during the Pleistocene.

In North America, the most common species are *P. sagittatum* and *P. arifolium,* which occur mainly in the eastern part of the continent (Fig. 8). Three other taxa, *P. perfoliatum, P. senticosum* var. *senticosum,* and *P. bungeanum,* were recently introduced from Asia, and have become established as successful weeds (see the taxonomic treatment for distributions).

Polygonum sagittatum is disjunct in the two widely separated regions of eastern Asia and eastern North America (Fig. 10). These disjuncts show weak morphological differentiation, but are not separable as species (see the taxonomic treatment). The mode of origin of these disjuncts has long been in dispute (Fernald, 1931; Hara, 1952; Li, 1972; Ohki, 1926; Steward, 1930).

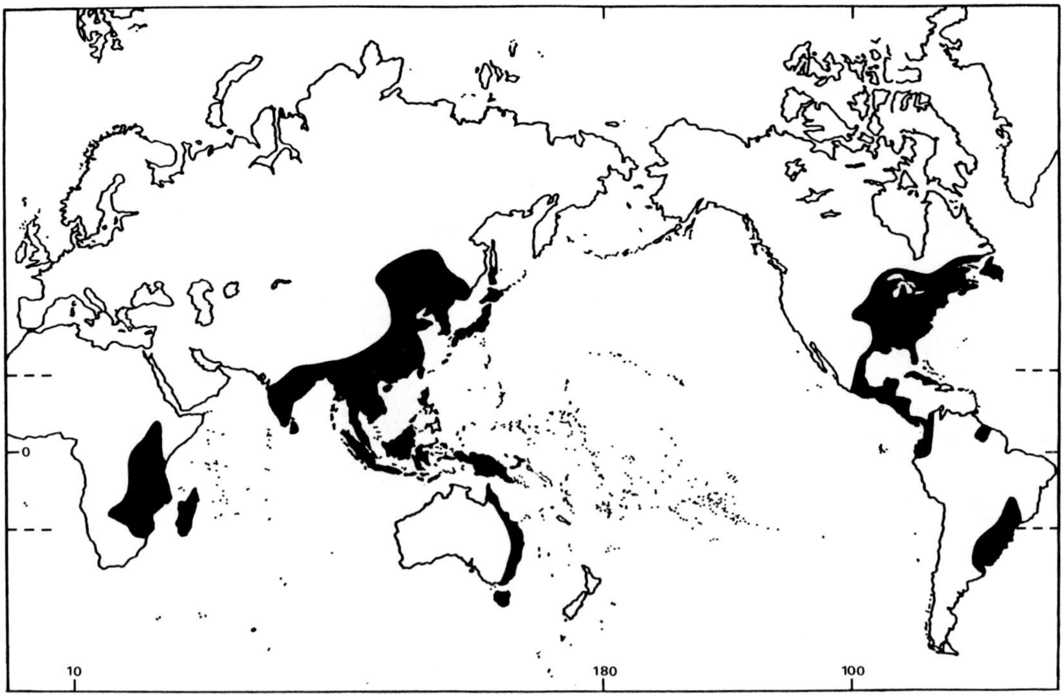

Fig. 6. Worldwide distribution of *Polygonum* section *Echinocaulon.*

Although paleobotanical data concerning *P. sagittatum* are lacking, some indirect evidence suggests that the species may once have had a more continuous range in the northern hemisphere, perhaps before the Pleistocene glaciation. As mentioned in the taxonomic treatment, North American populations are well within the range of variability of eastern Asiatic ones, and they show remarkable similarities in flavonoid chemistry in spite of their geographical remoteness (Park, 1987). Although parallel development due to similar environmental conditions cannot be ruled out, the occurrence of an intermediate form (form 3: see the taxonomic treatment) on both continents, predominantly in northern regions, may also suggest vicariant origin of these disjunct populations. Also, the present distribution pattern is strictly confined to eastern Asia and eastern North America, a fact suggestive of an Arcto-Tertiary relict disjunction pattern. The Arcto-Tertiary Geoflora and its history influenced the origin of eastern Asian-eastern North American disjuncts. A series of temperate forest associations were believed to be once widespread over high northern latitudes during the Eocene, but their fragments become separated into two disjunct regions, eastern Asia and eastern North America, during Pleistocene glaciation (Chaney, 1947; Graham, 1972). One line of evidence which may support this hypothesis comes from *P. arifolium. Polygonum arifolium* is phylogenetically distinct from its sympatric species, *P. sagittatum,* and its closest relative is found in Asia. *Polygonum arifolium* more closely resembles *P. biconvexum* of Asia than it does South American taxa in many morphological characters, including habit, leaf shape, pubescence, and floral morphology. Nevertheless, the similarity due to parallel development cannot be totally ruled out.

Another possible explanation, in this case, would be long-distance dispersal. Although the seeds of *P. sagittatum* do not seem suited for long-distance dispersal either by wind or by flotation, the seeds are frequently eaten by the far-ranging, widely distributed duck species, *Anas strepera* L., and possibly by other species of birds (Ridley, 1930). According to Ridley, this bird breeds in northern Europe, Asia, and North America and ranges south in winter to southern Asia, nearly to Africa, and as far south in America as the southern end of California, and as far east as China and Japan. If the establishment of

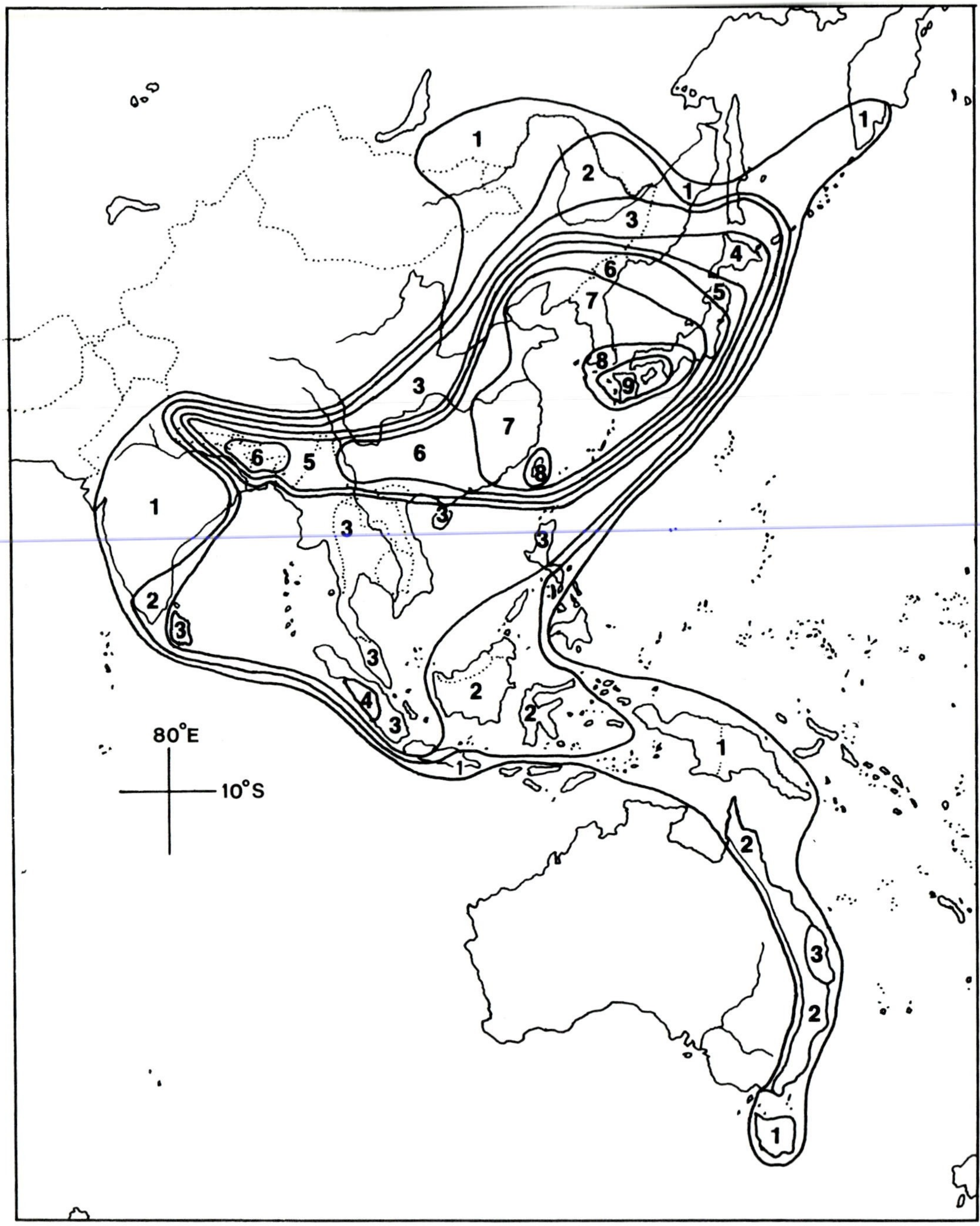

Fig. 7. Taxonomic diversity map of *Polygonum* section *Echinocaulon* in Asia and Australia. Species and varieties are counted.

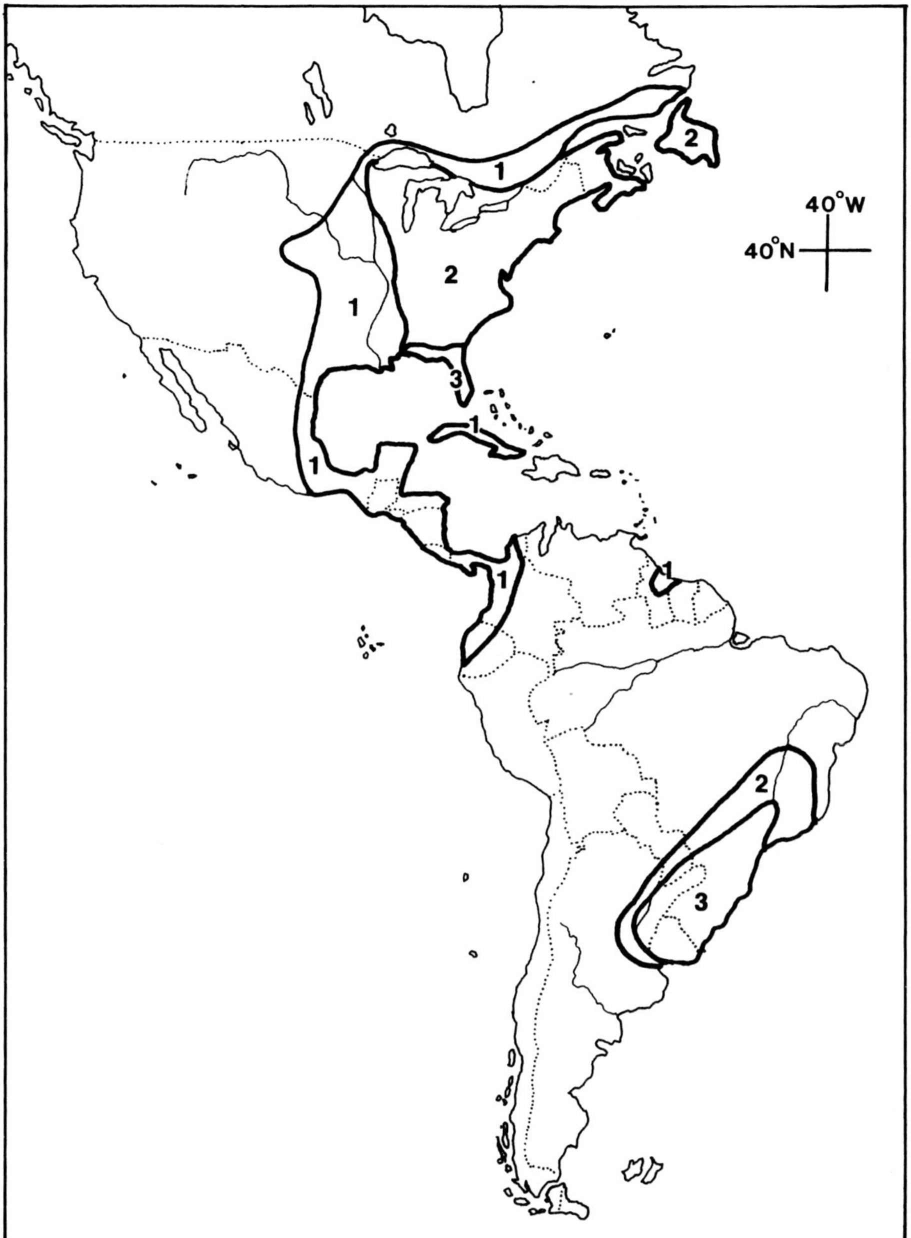

Fig. 8. Taxonomic diversity map of *Polygonum* section *Echinocaulon* in North and South America. Species and varieties are counted.

these disjunct populations was a result of long-distance dispersal by birds, it is surprising that this species is not found in southeast Asia, Africa, and western North America.

In South America, three species (*P. meisnerianum, P. stelligerum,* and *P. rubricaule*) are found, with the center of distribution in southern Brazil (Fig. 8). Although *P. meisnerianum* var. *beyrichianum* occurs in Mexico, Cuba, Colombia, and at least three states in the United States (Florida, Louisiana, and South Carolina), the South American species appear not to be closely

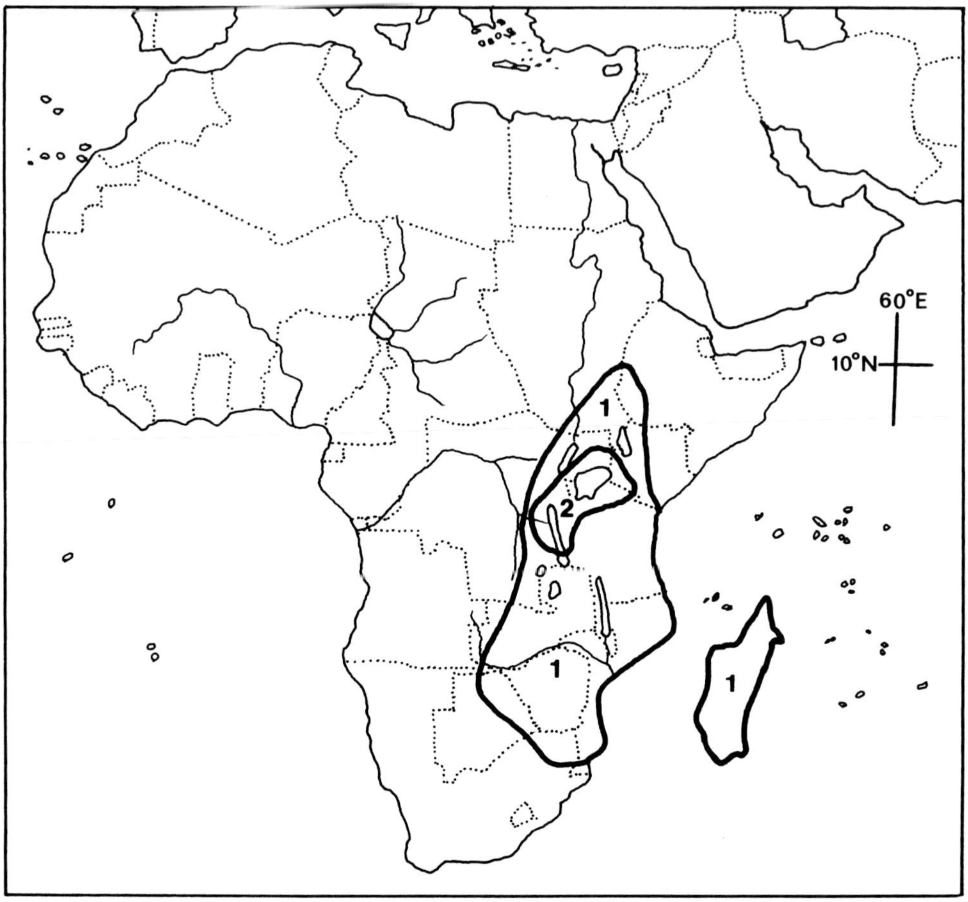

Fig. 9. Taxonomic diversity map of *Polygonum* section *Echinocaulon* in Africa. Species and varieties are counted.

related to either North American taxa or Asiatic ones. There is a much closer intercontinental relationship in distribution of the section between South America and southeastern Africa. Two taxa, *P. meisnerianum* var. *beyrichianum* and *P. rubricaule,* are shared by these two widely separated regions. The occurrence of common taxa between these two regions again presents problems concerning their origin and dispersal.

As in *P. sagittatum,* one possible explanation is that *P. meisnerianum* var. *beyrichianum* and *P. rubricaule* may represent the residual elements of once more or less continuous populations that have become isolated into widely separated populations. A close floristic relationship between South America and Africa has been suggested by several authors (Moore, 1973; Raven, 1972; Raven & Axelrod, 1974; Thorne, 1973), and at least 111 genera appear to be restricted to Africa (including Madagascar) and South America (Thorne, 1973). Raven and Axelrod (1974) suggested the importance of overland direct migration between South America and Africa to explain the fundamental floristic relationship between the two continents. According to Raven and Axelrod, the gradual isolation of Africa from South America by progressive sea-floor spreading occurred during the late Cretaceous Period, and overland direct migration had been possible until early Paleocene time. They also suggested that South America was more accessible to immigration from Africa than from North America until after the early Eocene. Unfortunately, there is no direct evidence, except the present distribution patterns of these taxa, to support overland direct migration during the Paleocene or age of this group on these two continents. However, existence of several endemic species of section

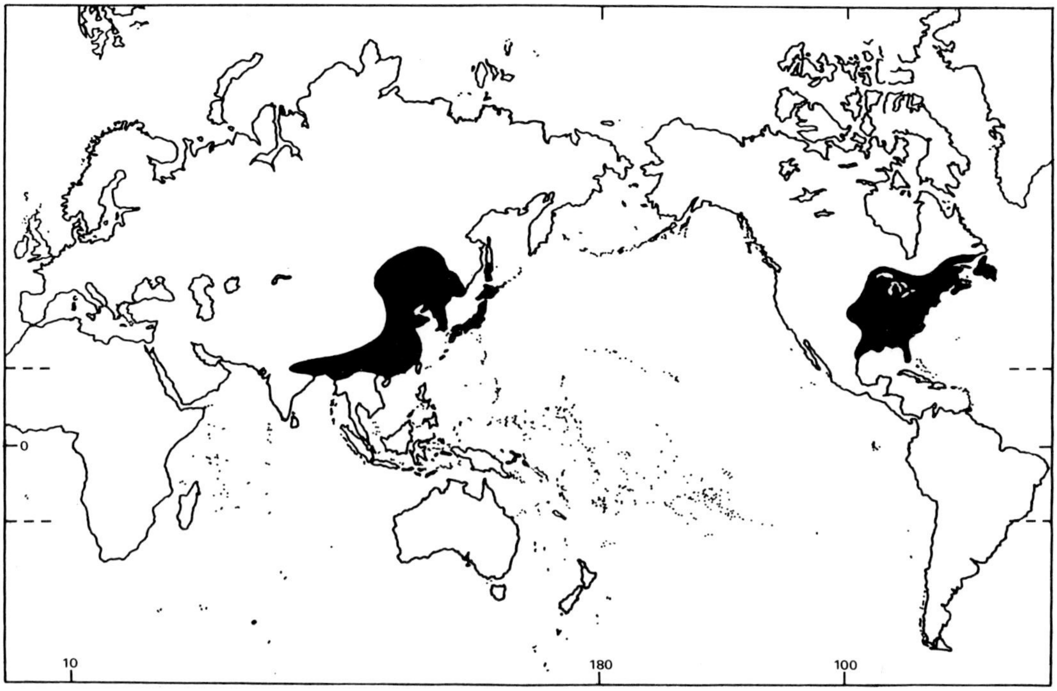

Fig. 10. Worldwide distribution of *P. sagittatum.*

Echinocaulon on each continent strongly suggests that the section has been present in Africa and South America for a considerable period of time.

Another possible explanation for the occurrence of common taxa between Africa and South America would be the relatively recent introduction of plants from one continent to the other. As pointed out earlier, seeds of *Polygonum* species do not seem suited for long-distance dispersal either by wind or by flotation. The only probable vectors that may be responsible for long-distance dispersal in this case are birds and humans. Long-distance dispersal by migrating birds cannot be ruled out, but the distances to be overcome are enormous. A more plausible explanation would be introduction by man. Points of circumstantial evidence suggesting humans as dispersal agents are: (1) some species in the section are frequently encountered in association with man, often as weeds in fields, roadsides, and other disturbed habitats; (2) these two areas were frequently connected by slave-trade ships during the 17th and 18th centuries (Curtin, 1969); and (3) present distribution pattern of *P. meisnerianum* var. *beyrichianum* in America (Fig. 11) parallels the major entry ports of slave ships during the 17th and 18th centuries. If the introduction of these species occurred recently by human dispersal, probably the greatest difficulty is to explain the origin of *P. stelligerum. Polygonum stelligerum* is morphologically distinct from other South American taxa, and no close relatives could be found either in North America or in Africa. Derivation from recently introduced species is highly unlikely. Alternatively, if we assume that introduction of these species occurred from South America to Africa by humans, explanation of the origin of other endemic species in Africa becomes problematical. However, dispersal by human agency should not be ruled out; clear examples of dispersal by human agency are the recent introductions of *P. perfoliatum, P. senticosum* var. *senticosum,* and *P. bungeanum* to North America from eastern Asia.

In summary, there are strong intercontinental relationships in the distribution of the section *Echinocaulon* between eastern Asia and eastern North America, and between South America and southeastern Africa. The section shows complex disjunct distribution patterns that presumably reflect historical episodes of vicariance and dis-

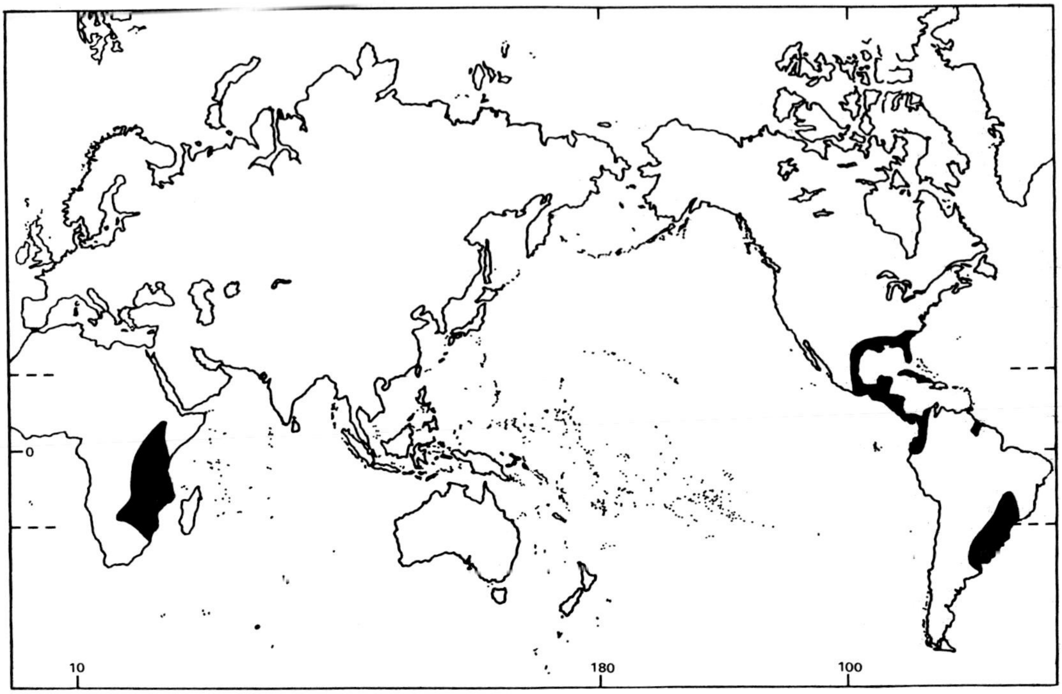

Fig. 11. Worldwide distribution of *P. meisnerianum* var. *beyrichianum*.

persal. However, possibilities of long-distance dispersals either by natural means or by humans cannot be ruled out.

Sectional Relationships and Phylogeny

Section *Echinocaulon* is a natural group in *Polygonum* characterized by the occurrence of recurved prickles on the angles of the stem, petioles, and abaxial surfaces of major leaf veins. Sharing of this unique character and a strong morphological resemblance among species in section *Echinocaulon* suggest that this group is monophyletic. Additional derived character states separating this group from other groups of *Polygonum* include scandent habit, the presence of stellate and tufted hairs (Haraldson, 1978), and elliptic to triangular leaf shapes, frequently with sagittate to hastate bases.

The section has traditionally been aligned with either of two others, *Persicaria* (Mill.) DC. or *Cephalophilon* Meisner. Although section *Cephalophilon* resembles some taxa of the section *Echinocaulon* in overall appearance, its phylogenetic relationship to section *Echinocaulon* is unclear. A close affinity between the two groups has been postulated by several authors (Dammer, 1892; Hedberg, 1946). Dammer (1892) recognized section *Echinocaulon* as a subsection of section *Cephalophilon* on the basis of similarities in the number of perianth lobes, the ocrea shape, the inflorescence, the leaf shape, and the cotyledon shape. However, most of the character states concerned also are shared by species of section *Persicaria* (Table III). A fundamental difference that seems to preclude a very close relationship between section *Echinocaulon* and section *Cephalophilon* is seen in the pollen morphology. Pollen grains of section *Cephalophilon* are substantially different from those of the sections *Echinocaulon* and *Persicaria*. The former group has tricolpate (rarely six- to eight-colpate) pollen grains, whereas the latter two have polyporate grains (Hara, 1966; Hedberg, 1946). Hedberg (1946) suggested a possible relationship between section *Cephalophilon* and section *Echinocaulon* based on similarities in exine structure of the pollen grains, and he considered *P. debile* to be intermediate between the two groups. But *P. debile* is clearly distinct from

Table III
Comparison of *Polygonum* section *Echinocaulon* with two possible allies

	Section *Echinocaulon*	Section *Persicaria*	Section *Cephalophilon*
Habit	usually scandent	ascending to erect	ascending to erect, often creeping
Stem	retrorsely prickly	unarmed	unarmed
Leaf shape	elliptic to triangular, rarely linear lanceolate	lanceolate to narrowly elliptic	broadly elliptic to ovate, often lobed
Leaf base	cordate to sagittate, often hastate	acute to attenuate	weakly cordate to truncate, often tapering to petiole
Petiole	rarely conspicuously winged	not winged	often winged, foliaceous bracts at the base
Ocrea	scarious or with foliaceous limb	scarious	scarious
Inflorescence	panicle- or head-like cluster	spikelike cluster	headlike cluster
Perianth lobes	4–5	4–5	4–5
Perigynous glandular teeth	present	usually present	absent
Cotyledon	accumbent	accumbent	accumbent
Pollen[a]	polyporate	polyporate	tricolpate

[a] Hara (1966), Hedberg (1946).

species of section *Cephalophilon* in having loose, interrupted paniclelike inflorescences, foliaceous ocreae, and entire, triangular leaves, characteristics that are not encountered in section *Cephalophilon.* Regardless of whether the two sections are related, it is doubtful that *P. debile* is an intermediate. Some additional differences between the two groups are listed in Table III.

A comparison of section *Echinocaulon* and section *Persicaria* suggests that section *Persicaria* is a more likely ally of section *Echinocaulon* (Table III). A close affinity between the two has been postulated by Gross (1913), and more recently by Hedberg (1946). Pollen grains of the two sections are indistinguishable (Hedberg, 1946), and some species in the section *Echinocaulon* closely resemble those of section *Persicaria.* In *P. bungeanum, P. dissitiflorum, P. praetermissum, P. rubricaule,* and *P. subsagittatum,* the bracts that subtend flower fascicles are funnelform and completely surround the inflorescence branch. This type of bract is characteristic of the section *Persicaria,* but is not encountered in section *Cephalophilon.* In addition, *P. bungeanum* closely approaches section *Persicaria* in several other morphological characters, and may be considered an intermediate taxon between the two sections. As in species of section *Persicaria,* its flower fascicles are arranged in relatively long, spikelike clusters, and leaves are lanceolate to narrowly elliptic with acute to attenuate bases. Komarov (1936) placed it in section *Persicaria,* but the presence of recurved prickles on the stem clearly aligns it with section *Echinocaulon.*

To provide an objective estimate of phylogeny in section *Echinocaulon,* cladistic analysis was employed. Cladograms were constructed using the PAUP program (version 2-4) of Swofford (1985), which is based on a parsimony procedure. Twenty-one morphological characters, which show the least intraspecific variation, were selected for the analysis (Table IV). The data matrix for section *Echinocaulon* is provided in Table V. Quantitative character states (characters C and J) were defined by discontinuities of variation, and their mean values are provided in Table I. Characters were polarized by outgroup comparison. Section *Persicaria,* which is most closely related to section *Echinocaulon,* was chosen as an outgroup. For each character, the most common state present in the outgroup was considered as a plesiomorphic character state for section *Echinocaulon* (Table IV). However, characters I and O to S were unordered because of the extreme morphological heterogeneity within both ingroup and outgroup. Multistate characters J and M also were unordered to avoid difficult assumptions of an a priori transforma-

Table IV

Characters and character states used in the cladistic analysis of *Polygonum* section *Echinocaulon*. Character state values are given in parentheses. Primitive character states are marked with an asterisk (*). Characters I, J, M and O to S were unordered

A. Habit: erect to ascending (0)*; climbing (1). **B.** Leaf shape: lanceolate to elliptic (0)*; triangular (1). **C.** Leaf length/width ratio: greater than 2 (0)*; less than 2 (1). **D.** Leaf base: acute to attenuate (0)*; cordate to sagittate or hastate (1). **E.** Petioles: not conspicuously winged (0)*; conspicuously winged (1). **F.** Stellate hairs: absent (0)*; present (1). **G.** Tufted hairs: absent (0)*; present (1). **H.** Simple unicellular hairs: absent (0)*; present (1). **I.** Glandular hairs (inflorescence branch): absent (0); present (1). **J.** Petiole length/leaf length ratio: almost sessile (0); 0.5–0.2 (1); greater than 0.4 (2). **K.** Ocrea shape: scarious (0)*; with foliaceous limb (1). **L.** Inflorescences: branched (0)*; unbranched (1). **M.** Inflorescence shape: loose interrupted panicle of 3–6 flower fascicles (0); intermediate (1); dense panicle or a headlike cluster of many flower fascicles (2). **N.** Bracts: funnelform and completely surrounding the inflorescence branch (0)*; elliptic and not surrounding the inflorescence branch (1). **O.** Flower fascicles: widely spaced (0); closely spaced (1). **P.** Perianth: 5-parted (0); 4-parted (1). **Q.** Stamen insertion: in 2 whorls (0); in 1 whorl (1). **R.** Style: 2-cleft (0); 3-cleft (1). **S.** Pedicels: included (0); completely exserted (1). **T.** Perianth: not becoming fleshy (0)*; becoming fleshy (1). **U.** Recurved prickles: absent (0)*; present (1).

tion series. The cladograms generated were rooted using the outgroup rooting method of Swofford (1985).

Figure 12 shows one of several most parsimonious cladograms, based on 21 morphological characters. It has a total length of 47 evolutionary steps and a consistency index (Kluge & Farris, 1969) of 0.489. Other equally parsimonious cladograms have similar topologies and differ mainly in certain branch lengths. *Polygonum strigosum* has been excluded from the analysis, because extreme polymorphism prevents scoring many important characters (see the taxonomic treatment).

The most important point to observe about patterns of character evolution within section *Echinocaulon* is the extent of homoplasy. Most of the characters that are useful and diagnostic in delimiting species appear to be homoplasious (Fig. 12). Many of the reproductive characters, which might be expected to show relative evolutionary stability, are incongruent with other characters and with each other. For instance, three reproductive characters (P, Q, R: parenthetical letters here and subsequently refer to character codes in Table IV) show extensive parallelisms and reversals on the cladogram (Fig. 12). The four-parted perianth condition (P:1) is derived two times (with one reversal), the single-whorled stamen arrangement (Q:1) is derived two times independently, and the three-cleft style (R:1) evolved three times (with three reversals). These characters all have conflicting distributions (Fig. 12).

Of the 18 non-autapomorphic characters, only six characters (A, B, D, E, K, U) do not show parallelism or reversal; two of these characters (D, U) accounted for changes between the outgroup and ingroup taxa. Triangular leaf shape (B:1) has arisen only once in the cladogram, and defines taxa 1 to 9 (taxon numbers correspond to those in Fig. 12) as a distinct lineage within the section. Although this lineage is defined by only two synapomorphies (B:1, P:1), these taxa were consistently clustered together in all cladograms generated during the study. The distribution of triangular leaf shape in the section corresponds with that of elongate petioles (Fig. 12). The high petiole length/leaf length ratio (J:2) is shared by taxa 1 to 8. The only exception is *P. stelligerum* (taxon 9). The mean petiole length/leaf length ratio of *P. stelligerum* is 0.31 (range: 0.2–0.43), and is intermediate between taxa 1 to 8 and the remaining taxa (taxa 10 to 24); this is due to a much longer leaf length relative to leaf width in *P. stelligerum* (Table I). Likewise, the derivation of linear lanceolate leaves (J:0) in *P. meisnerianum* var. *beyrichianum* (taxon 20) and *P. dichotomum* var. *angustissimum* (taxon 22) corresponds with reduction in petiole length (Fig. 12).

Derivation of the foliaceous ocrea (K:1) also seems to parallel evolution of triangular leaf shape. The foliaceous ocrea is shared by taxa 1 to 7, all of which have triangular leaves. But *P. arifolium* (taxon 8) and *P. stelligerum,* which also have triangular leaves, have scarious, oblique ocreae. Ocreae of *P. arifolium* are very short and somewhat funnelform, nearly approaching those of *P. debile* (taxon 4) and *P. biconvexum* (taxon 7) in overall appearance (Fig. 3). Although *P. arifolium* is endemic to eastern North America and is far separated geographically from these eastern Asian taxa 1 to 7, it closely resembles *P.*

Table V

Data matrix for cladistic analysis of *Polygonum* section *Echinocaulon*. Character numbers and character states correspond to those in Table IV. Species are designated by the first three letters of their epithets. SEN-1: var. *senticosum*. SEN-2: var. *sagittifolium*. MEI-1: var. *meisnerianum*. MEI-2: var. *beyrichianum*. THU-1: var. *thunbergii*. THU-2: var. *maackianum*. DIC-1: var. *dichotomum*. DIC-2: var. *angustissimum*. Outgroup; section *Persicaria*

	Characters																				
Taxon	A	B	C	D	E	F	G	H	I	J	K	L	M	N	O	P	Q	R	S	T	U
PER	1	1	1	1	0	0	0	0	0	2	1	1	2	1	1	0	0	1	0	1	1
SEN-1	1	1	1	1	0	0	0	1	1	2	1	0	2	1	1	0	0	1	0	0	1
SEN-2	1	1	0	1	0	0	0	0	0	2	1	0	2	1	1	0	0	1	0	0	1
ARI	0	1	1	1	0	1	0	0	1	2	0	0	0	1	0	1	0	0	0	0	1
DEB	0	1	1	1	0	1	0	0	1	2	1	0	0	1	0	0	0	1	0	0	1
BRE	0	0	0	1	0	1	0	0	1	1	0	0	0	1	0	0	0	1	0	0	1
DIS	0	0	1	1	0	1	0	0	1	1	0	0	0	0	0	0	0	1	0	0	1
PRA	0	0	0	1	0	0	0	0	1	1	0	0	0	0	0	1	1	1	0	0	1
STE	0	1	0	1	0	1	0	0	1	1	0	0	2	1	1	1	0	0	0	0	1
BUN	0	0	0	1	0	0	0	0	1	1	0	0	0	0	0	0	0	0	0	0	1
MUR	0	0	0	1	0	1	1	1	1	1	0	0	2	1	1	0	0	1	0	0	1
HAS	0	0	0	1	0	1	1	0	1	1	0	0	2	1	1	0	0	1	1	0	1
MEI-1	0	0	0	1	0	0	1	0	1	1	0	0	1	1	1	0	1	1	0	0	1
MEI-2	0	0	0	1	0	0	1	0	1	0	0	0	1	1	1	0	1	1	0	0	1
SUB	0	0	0	1	0	0	1	0	1	1	0	0	0	0	0	0	1	0	0	0	1
RUB	0	0	0	1	0	0	0	0	0	1	0	0	0	0	0	0	0	1	0	0	1
THU-1	0	1	1	1	1	1	0	0	1	2	1	0	2	1	1	0	0	1	0	0	1
THU-2	0	1	1	1	1	1	0	0	1	2	1	0	2	1	1	0	0	1	0	0	1
BIC	0	1	1	1	0	1	0	0	1	2	1	0	2	1	1	1	0	0	0	0	1
BRA	0	0	0	1	0	1	0	0	1	1	0	0	1	1	1	0	1	0	0	0	1
CLA	0	0	0	1	0	0	0	0	1	1	0	0	1	1	1	0	1	1	0	0	1
SAG	0	0	0	1	0	0	0	0	0	1	0	0	2	1	1	0	0	1	0	0	1
DIC-1	0	0	0	1	0	0	0	0	1	1	0	0	1	1	1	0	1	0	0	0	1
DIC-2	0	0	0	1	0	0	0	0	1	0	0	0	1	1	1	0	1	0	0	0	1
OUTGROUP	0	0	0	0	0	0	0	0	1	1	0	0	2	0	1	0	0	0	0	0	0

biconvexum in many morphological characters, including habit, leaf shape, pubescence, and floral morphology. In this respect, *P. arifolium* may be a possible link between *P. stelligerum* and taxa 1 to 7 in terms of ocrea shape and texture, but phylogenetic relationship of *P. stelligerum* to taxa 1 to 7 is unclear (see the taxonomic treatment).

Divergence of the remaining taxa (taxa 10 to 24) is much more complicated by extensive parallelisms (Fig. 12). Most of the lower nodes on the cladogram are defined by character states that show many parallelisms and reversals, yet several basic lineages may still be recognized. The first lineage includes two sympatric species, *P. muricatum* (taxon 10) and *P. hastatosagittatum* (taxon 11). Although this lineage is defined by only one character (G) that shows many parallelisms, the two species are very similar in most morphological characters, including those not used in this analysis, and in flavonoid chemistry (Park, 1987). The second lineage, defined by two inflorescence characters (M, O), comprises taxa 12 to 17. Within this lineage, *P. breviochreatum* (taxon 12) appeared to be distant from the remaining taxa. Unlike other taxa in the lineage, *P. breviochreatum* has broadly elliptic bracts that do not surround the inflorescence branch (N:1). In addition, its distribution is highly restricted to southern Japan and southern Korea. It also has a zero branch length, suggesting that the relationship shown on the cladogram may be due to ignorance of potentially resolving characters.

Polygonum sagittatum (taxon 18) appeared as a phylogenetically isolated species in the section (Fig. 12). Its generated position in the various cladograms was unstable and always distant from other taxa. The same relationship has been suggested previously on the basis of comparative flavonoid chemistry (Park, 1987). The last lineage includes taxa 19 to 24, a group of morpho-

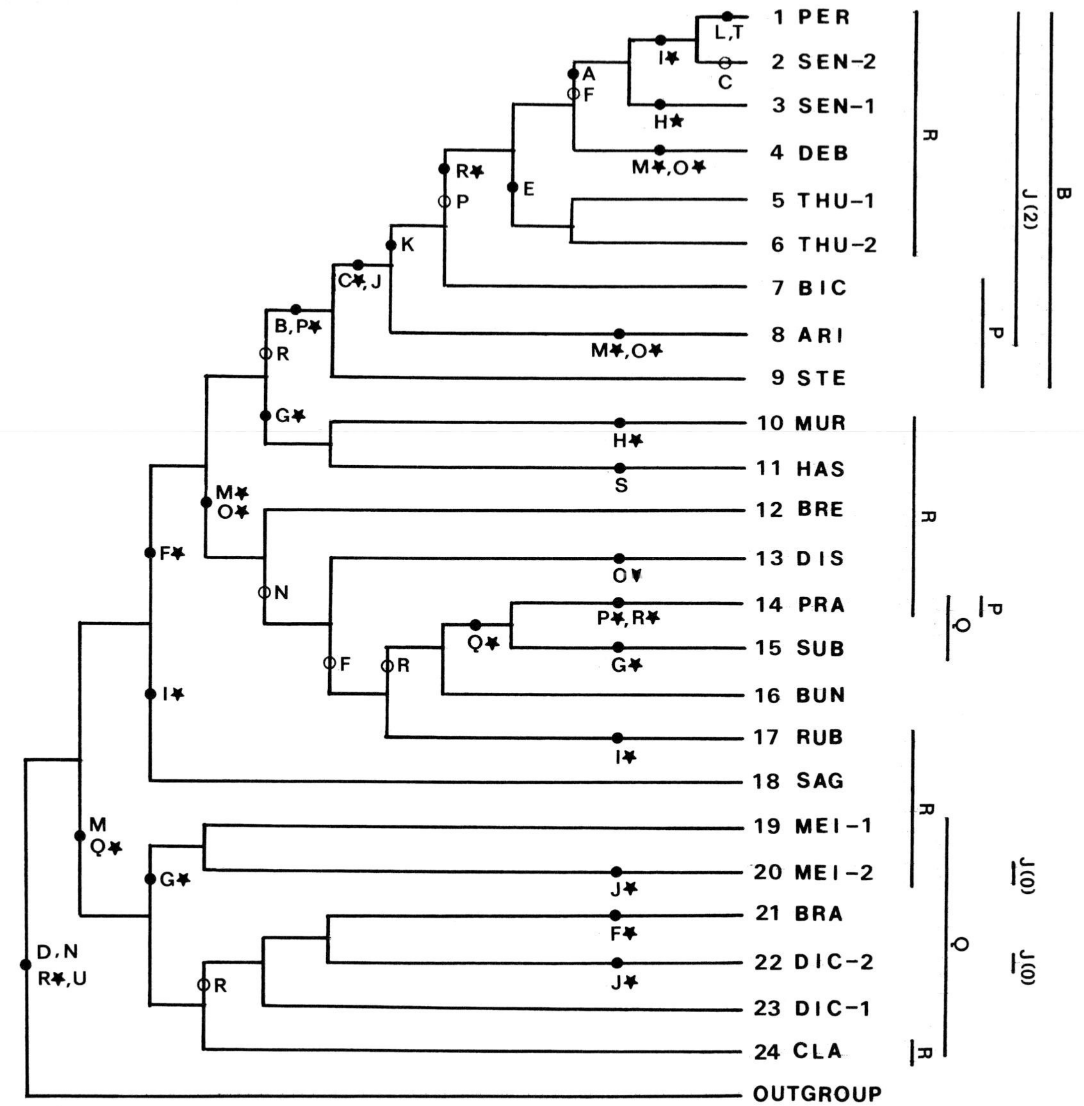

Fig. 12. Cladogram of species of *Polygonum* section *Echinocaulon* based on 21 morphological characters. Species are designated by the first three letters of their epithets. Closed circles = apomorphies, stars = parallelisms, open circles = reversals, solid bars = derived states. Character codes correspond to those in Table IV.

logically variable taxa, and it is defined by inflorescence shape (M:1) and stamen arrangement (Q:1). Divergence of these taxa within the lineage is not fully resolved and their relationships are not clear, but these taxa were consistently clustered together in all cladograms.

Nomenclatural History

A summary of the nomenclatural history of *Polygonum* section *Echinocaulon* was provided by Park (1986b). In my studies, I have chosen to follow the generic concept and delimitation of Steward (1930) and Graham and Wood (1965), who recognized the group as a section of the genus *Polygonum.*

Taxonomic Treatment

Polygonum sect. **Echinocaulon** Meisner *in* Wallich, Pl. Asiaticae rar. **3:** 58. 1832. *Echinocaulon* (Meisner) Spach, Hist. nat. vég. **10:** 521.

1841. *Echinocaulon* (Meisner) Hasskarl, Flora **25(2):** 20. 1842 ("*Echinocaulos*"). *Polygonum* sect. *Cephalophilon* subsect. *Echinocaulon* (Meisner) Dammer *in* Engler & Prantl, Nat. Pflanzenfam. **3(1a):** 28. 1892. *Polygonum* subgen. *Echinocaulon* (Meisner) Small, Mem. Dept. Bot. Columbia Coll. **1:** 26. 1895. *Persicaria* sect. *Echinocaulon* (Meisner) Gross, Bull. Acad. Int. Géogr. Bot. **23:** 27. 1913. Lectotype. *P. muricatum* Meisner, designated by Park, Brittonia **38:** 396. 1986.

Truellum Houttuyn, Natuurlyke historie **8:** 427. 1777. *Truellum* sect. *Truellum,* Sojak, Preslia **46:** 144. 1974. Type. *T. japonicum* Houttuyn (=*Polygonum senticosum* (Meisner) Franchet & Savat.).

Helxine subgen. *Belotrachis* Raf., Fl. tellur. **3:** 11. 1837. Type. *H. sagittatum* (L.) Raf. (≡*Polygonum sagittatum* L.).

Tracaulon Raf., Fl. tellur. **3:** 13. 1837. *Polygonum* sect. *Tracaulon* (Raf.) Correll & Johnston, Manual Vasc. Pl. Texas 520. 1970, nom. illeg. Type. *T. arifolium* (L.) Raf. (≡*Polygonum arifolium* L.).

Helxine subgen. *Tasoba* Raf., Fl. tellur. **3:** 94. 1837. Lectotype. *H. arifolia* Raf. (≡*Polygonum thunbergii* Siebold & Zucc.), designated by Park, Brittonia **38:** 396. 1986.

Chylocalyx Hasskarl ex Miquel, Fl. Ned. Ind. **1(1):** 1012. 1858. *Persicaria* sect. *Chylocalyx* (Hasskarl ex Miquel) Nakai, Rigakkai **24:** 298. 1926. Type. *C. perfoliatus* (L.) Hasskarl ex Miquel (≡*Polygonum perfoliatum* L.).

Persicaria sect. *Dissitiflorae* Nakai, Rigakkai **24:** 297. 1926. Type. *P. dissitiflora* (Hemsley) Gross ex Mori (≡*Polygonum dissitiflorum* Hemsley).

Ascending to scandent or climbing herbs, 0.3–4 m tall, rarely up to 6 m tall; stems usually branched, sparsely to densely retrorsely prickly along the angles; leaves alternate, simple, entire, subsessile to long-petioled; blades triangular or linear lanceolate to ovate, acute to acuminate at the apex, at the base weakly cordate to sagittate, often truncate to hastate, prickly along major veins beneath; ocreae foliaceous or scarious; inflorescences terminal or in the axils of upper leaves, simple or branched, usually elongate; branches of inflorescence bracteate at the base; inflorescence bracts resembling ocreae but much reduced; each inflorescence branch terminated by a somewhat spikelike or headlike cluster of flower fascicles; flower fascicles usually 2- and 4-flowered, each subtended by a bract; bracts usually elliptic; flowers bisexual, regular, each bearing bracteoles; bracteoles paired, scarious, fused at the base; pedicels very short, usually included in bracteoles, rarely exserted, articulated at the perianth base; perianth petaloid, whitish-green to reddish-purple, 4- or 5-parted; tepals elliptic to broadly elliptic, persistent, enclosing the achene at maturity, rarely becoming fleshy; stamens 4–8, included, alternating with small flattened glandular teeth, inserted at the base of perianth in one or two whorls; filaments filiform, glabrous; anthers 2-locular, longitudinally dehiscent; ovary one, unilocular, 2- or 3-angled; style one, 2- or 3-cleft, glabrous, included, rarely slightly exserted; stigmas 2–3, minutely capitate; achenes biconvex or trigonous, usually dark brown to black, rarely yellowish-brown, smooth or rugose, shiny or dull, stipitate at the base.

In a previous publication (Park, 1986b), the lectotypification of section *Echinocaulon* by Roberty and Vautier (1964) was overlooked. However, *P. sagittatum,* designated by Roberty and Vautier, has been the type of *Helxine* subgenus *Belotrachis* Rafinesque.

Key to the Species of section *Echinocaulon*

1. Ocreae, at least some of them, foliaceous or with a marginal foliaceous limb.
    2. Inflorescences unbranched; perianth becoming fleshy and blue in fruit. — 1. *P. perfoliatum.*
    2. Inflorescences branched; perianth not becoming fleshy or blue in fruit.
        3. Leaf blades usually truncate to weakly cordate at the base.
            4. Plants stout, 1–2.5 m tall, climbing; each inflorescence branch terminated by a dense, headlike cluster of many flower fascicles; achenes dull. — 2. *P. senticosum.*
            4. Plants slender, 30–50 cm tall; each inflorescence branch terminated by a loose, interrupted panicle of 2–3 flower fascicles; achenes shiny. — 3. *P. debile.*
        3. Leaf blades conspicuously hastate with 2 divergent triangular basal lobes.
            5. Perianth 5-parted; achenes trigonous. — 4. *P. thunbergii.*
            5. Perianth 4-parted; achenes biconvex. — 5. *P. biconvexum.*
1. Ocreae completely scarious.
    6. Bracts subtending flower fascicles elliptic to broadly elliptic, not surrounding the inflorescence branch.
        7. Ocreae oblique at the apex.
            8. Leaf blades triangular in outline; perianth 4-parted.
                9. Leaf blades broadly triangular, 6.5–13 cm long, 6–10.8 cm wide; ocreae 0.5–1 cm long,

usually glabrous; each inflorescence branch terminated by a loose, interrupted panicle of 3–8 flower fascicles; plants of eastern North America. 6. *P. arifolium.*

9. Leaf blades narrowly triangular, 9.5–20 cm long, 5.5–14 cm wide; ocreae 2–4.5 cm long, densely stellate-pubescent; each inflorescence branch terminated by a dense, headlike cluster of many flower fascicles; plants of South America. 7. *P. stelligerum.*

8. Leaf blades lanceolate to narrowly elliptic; perianth 5-parted.
10. Inflorescence branches glabrous; stamens 8, inserted at the base of perianth in 2 whorls. 17. *P. sagittatum.*
10. Inflorescence branches glandular-pubescent; stamens usually 5, in 1 whorl.
11. Style 3-cleft; achenes trigonous.
12. Ocreae short-ciliate; plants of South America. 18. *P. meisnerianum.*
12. Ocreae eciliate; plants of India. 21. *P. clarkei.*
11. Style 2-cleft; achenes biconvex.
13. Leaf blades usually hastate with 2 acute triangular basal lobes; ocreae short-ciliate; plants of Madagascar. 19. *P. brachypodum.*
13. Leaf blades usually attenuate at the base; ocreae rarely ciliate; plants of northern India to Australia through southeast Asia. 20. *P. dichotomum.*

7. Ocreae truncate at the apex.
14. Ocreae short, 2–5 mm long, densely hispid with long appressed bristles; each inflorescence branch terminated by a loose, interrupted panicle of 2–4 flower fascicles. 11. *P. breviochreatum.*
14. Ocreae long, cylindrical, 1–3.5 cm long, long-ciliate, glabrous or bearing a few bristles along the veins; each inflorescence branch terminated by a relatively dense panicle or a headlike cluster of many flower fascicles.
15. Ocreae bearing bristles at least near the base. 10. *P. strigosum.*
15. Ocreae glabrous.
16. Leaf blades usually ovate to elliptic-oblong, weakly cordate at the base; pedicels shorter than bracteoles, included, glabrous. 8. *P. muricatum.*
16. Leaf blades usually lanceolate to narrowly elliptic, weakly sagittate with small rounded or triangular basal lobes; pedicels longer than bracteoles, exserted, glandular-pubescent. 9. *P. hastatosagittatum.*

6. Bracts subtending flower fascicles funnelform, completely surrounding the inflorescence branch.
17. Leaf blades acute at the base; ocreae truncate at the apex. 15. *P. bungeanum.*
17. Leaf blades usually sagittate to weakly hastate at the base; ocreae oblique at the apex.
18. Perianth 4-parted. 13. *P. praetermissum.*
18. Perianth 5-parted.
19. Ocreae eciliate, glabrous; inflorescence branches glabrous. 16. *P. rubricaule.*
19. Ocreae short-ciliate at the tip, often hispid; inflorescence branches glandular-pubescent.
20. Leaf blades large, broadly elliptic to ovate, 4.5–15 cm long, 2–9.5 cm wide; stamens inserted at the base of perianth in 2 whorls; achenes trigonous; plants of eastern Asia. 12. *P. dissitiflorum.*
20. Leaf blades linear lanceolate to lanceolate, 6.5–10 cm long, 0.8–1.7 cm wide; stamens in 1 whorl; achenes biconvex; plants of Africa. 14. *P. subsagittatum.*

1. Polygonum perfoliatum L., Syst. nat., ed. 10. 1006. 1759. *Fagopyrum perfoliatum* (L.) Raf., Fl. tellur. **3:** 10. 1837. *Echinocaulon perfoliatum* (L.) Meisner ex Hasskarl, Flora **25(2):** 20. 1842 ("*Echinocaulos perfoliatus*"). *Chylocalyx perfoliatus* (L.) Hasskarl ex Miquel, Fl. Ned. Ind. **1(1):** 1012. 1858. *Tracaulon perfoliatum* (L.) Greene, Leafl. Bot. Observ. Crit. **1:** 22. 1904. *Persicaria perfoliata* (L.) Gross *in* Loesener, Beih. Bot. Centralbl. **37:** 113. 1920. *Ampelygonum perfoliatum* (L.) Roberty & Vautier, Boissiera **10:** 31. 1964. *Truellum perfoliatum* (L.) Sojak, Preslia **46:** 148. 1974. Type. India. Without specific locality and date, *Burman s.n.* [holotype, LINN 510.35 (microfiche at NY!)].

Scandent or climbing glabrous *herb,* somewhat woody at the base, 1–6 m tall; stem green, branched, angled, densely armed with recurved prickles along the angles; prickles stout, broad-based, ca. 1 mm long. *Leaves* long-petioled; blades thin, membranous, pale green, often glaucous below, triangular, 4–7 cm long, 4.5–9 cm wide, acute to subacute at the apex, truncate to weakly cordate, often weakly hastate with obtuse triangular lobes at the base, usually sparsely retrorsely prickly at the margin; midveins and major veins beneath retrorsely prickly; petioles 4.5–8 cm long, moderately armed with recurved prickles. *Ocreae* foliaceous, green, orbicular, perfoliate, completely surrounding the node, 0.9–1.4 cm in diameter. *Inflorescences* terminal or in the axils of upper leaves, elongate, unbranched, 5–12 cm long, retrorsely prickly; inflorescence bracts resembling ocreae; each inflorescence axis terminated by a dense somewhat spikelike cluster of 10–15 flower fascicles. *Flower fascicles* 2- to 3-flowered, closely spaced; bracts elliptic, acuminate, 1–2 mm long; bracteoles broadly elliptic, ca. 2 mm long; pedicels very short, 2–3 mm long, almost included in bracteoles; perianth pale greenish-white, deeply 5-parted, 2–3.5 mm long; tepals broadly elliptic, becoming fleshy and blue in fruit; stamens eight, inserted at the base of perianth in two whorls; inner stamens three, ca. 1.2 mm long; outer stamens five, ca. 1 mm long; style one, ca. 1.5 mm long, deeply 3-cleft, included; stigmas three. *Achenes* spheroidal, black, smooth, shiny, 3–3.5 mm long.

Distribution. Predominantly in moist situations, usually in shade; occurring naturally in open thickets and along streambanks; common as a weed in abandoned fields, pastures, roadsides, and other disturbed sites; extending from Japan and Korea, through eastern Siberia, China, Taiwan, northern Philippines and Indochina, to northern India and Indonesia; recently introduced and becoming a successful weed in eastern United States (Pennsylvania and Maryland) and Canada (Fig. 13).

Polygonum perfoliatum is closely related to *P. senticosum,* and the two species are very similar in most morphological characters including an extensive climbing habit, long-petioled triangular leaves, and foliaceous ocreae. The two species also have identical flavonoid profiles (Park, 1987), providing additional evidence of close phylogenetic relationship. *Polygonum perfoliatum* can be readily distinguished from *P. senticosum* by its unbranched inflorescence and fleshy texture of perianth in fruit.

Additional specimens examined. BURMA. Mali Valley, S of Putao, 2000 ft, 12 Jan 1914, *Ward s.n.* (E).

CANADA. British Columbia: 2½ mi N and 1 mi E of Pitt Meadows, patches growing over sprayed Canada Thistle by road, 8 Sep 1954, *Faris 261* (DAO).

CHINA. Fukien: Diongloh and vic., 24 Jul 1926, *Lin 114432* (MSC); Foochow, Fukien Christian University Campus, 1 Jun 1925, *Metcalf & his class 965* (MT); Foochow and vic., 26 Aug 1924, *Metcalf & Tang 450* (MT); Yang Chi Hsing, 15 Apr 1937, *Migo s.n.* (PE); Foochow, Kuai Ang, houseside, damp place, 8 Nov 1926, *Tang 5565* (MT); Kushan Monastery, 21 Aug 1926, *Uong 12413* (MSC). Hopeh: Without specific locality, in 1913, *Clemens 8054* (E); Changli, 23 Jun 1930, *Hsia 1951* (NY); Peking, in 1856, *Tatarinow s.n.* (LE). Hunan: I-Chia-Ao, Changning Hsien, 26 Jun 1935, *Fan & Li 58* (LE). Hupeh: En Shin Hsien, 20 Jun 1934, *Chow 2001* (NY); Wuchang, Lokiashan, 22 Sep 1933, *Sun 780* (NY); Wuchang, Lokiashan, 30 Sep 1933, *Sun 852* (NY); west Hupeh, without date, *Wilson 2773* (E, NY). Kiangsi: Without specific locality, 1868–1873, *David s.n.* (P); Sungwu, 410 m, Jun–Jul 1936, *Gressitt 1636* (E); Kuling, Lu-shan, damp ground, 8 Jul 1922, *Steward 2463* (MT); vic. Lingchuan, in open waste places, 115 m, 15 Jun 1932, *Tsiang 9791* (NY). Kwangtung: Hainan, Po-ting, 2000 ft, 3 Apr 1935, *How 71595* (F); Wung-yuen Dist., Yung-yun City and vic., 22 Dec 1932, *Lau 851* (NY); Hainan, Ching Mai Dist., Pak Shik Ling and vic., Ku Tung Village, 20 Apr 1933, *Lei 561* (F, NY); Hainan, 12 Jan 1934, *Liang 64490* (NY); Honan Isl., 10 Apr 1917, *Levine 495* (F); Hainan, Taam-chau Dist., ShanTong To, Sha Po Shan and vic., 19 Aug 1927, *Tsang & Lu 16007* (NY); Yingtak Dist., in valley, 31 Jul 1929, *Wang 2905* (F, NY). Kweicho: Without specific locality, 750–1000 m, 11 Jul 1917, *Handel-Mazzetti 2102* (P). Liaoning: Antung, Jul 1928, *Makino 16617* (MAK). Shantung: Tai-shan Mts., Sep 1912, *Clemens 1366* (E). Szechwan: Without specific locality, in 1893, *Potanin s.n.* (LE). Yunnan: Without specific locality, 2 May 1890, *Delavay s.n.* (NY); without specific locality and date, *Drake 3202* (L); Yung-peh Mts., Sep 1918, *Forrest 16922* (E); Tihen-fang-shan, in 1921, *Maire 272* (P).

HONG KONG. Without specific locality and date, *Ford s.n.* (NY); Shek-O Rd., 21 Oct 1950, *Tang 667* (FLAS).

INDIA. Assam: Khasi Hills, Shillong, 6000 ft, 9 Sep 1954, *Chand 8147* (L); Kalain, Cachar, 9 Dec 1885, *Clarke s.n.* (NY); "Mont. Khasia, Regio. Trop.," without date, *Hook. f. & Thomson s.n.* (NY); Shillong, Jawai, on the rd. to the E of the JB., 28 Oct 1956, *Panigrahi 4104* (L). Manipur: Karong, 3500 ft, 24–30 Sep 1950, *Chand 3715* (L); Ukhrul, 6000 ft, 22 Aug 1948, *Kingdom-Ward 17978* (NY).

INDONESIA. Java: Res. Preanger, 1200 m, in 1909, *Backer 31965* (L); Semarang, 25 Jun 1897, *Koorders 28044B* (L); Tjibadas, Preanger, ca. 1000 m, 16 Oct 1898, *Koorders 31721B* (L); Tjikadjang, 27 May 1875, *Kuntze 4588* (NY). Sumatra: Without specific locality, in 1864, *Korthals 161* (NY); Sibalangit, NW rand. Toba, ca. 1400 m, 30 Jan 1920, *Larring 1252* (L); Karo Plateau, nr. Kabandjahe, 1100–1250 m, *Lörzing 13717* (L); Toba, ca. 1250 m, 14 Jul 1897, *Nuwcharst? 356* (L).

JAPAN. Without specific locality, 17 Sep 1887, *Faurie 1160* (P). Aomori: Shimokita-gun, Obatake-cho, 18 Aug 1957, *Mori s.n.* (MAK 22996); Aomori-shi, Komagome, 15 Aug 1971, *Naito s.n.* (TUS). Chiba: Higashikatsuura-gun, Nagareyama-cho, 30 Sep 1904, *Ishihara s.n.* (MAK 16627). Fukushima: Nishikawa-gun, Jul 1907, *Imai s.n.* (MAK 16620). Ibaragi: Namegata-gun, Itako-cho, 17 Sep 1901, *Suzuki s.n.* (MAK 16621). Kanagawa: Yokohama, 13 Oct 1862, *Maximowicz s.n.* (LE). Kochi: Tosa, in 1905, *Makino 16632* (MAK). Kyoto: Prov. Yamashiro, Katsura, W of Kyo-

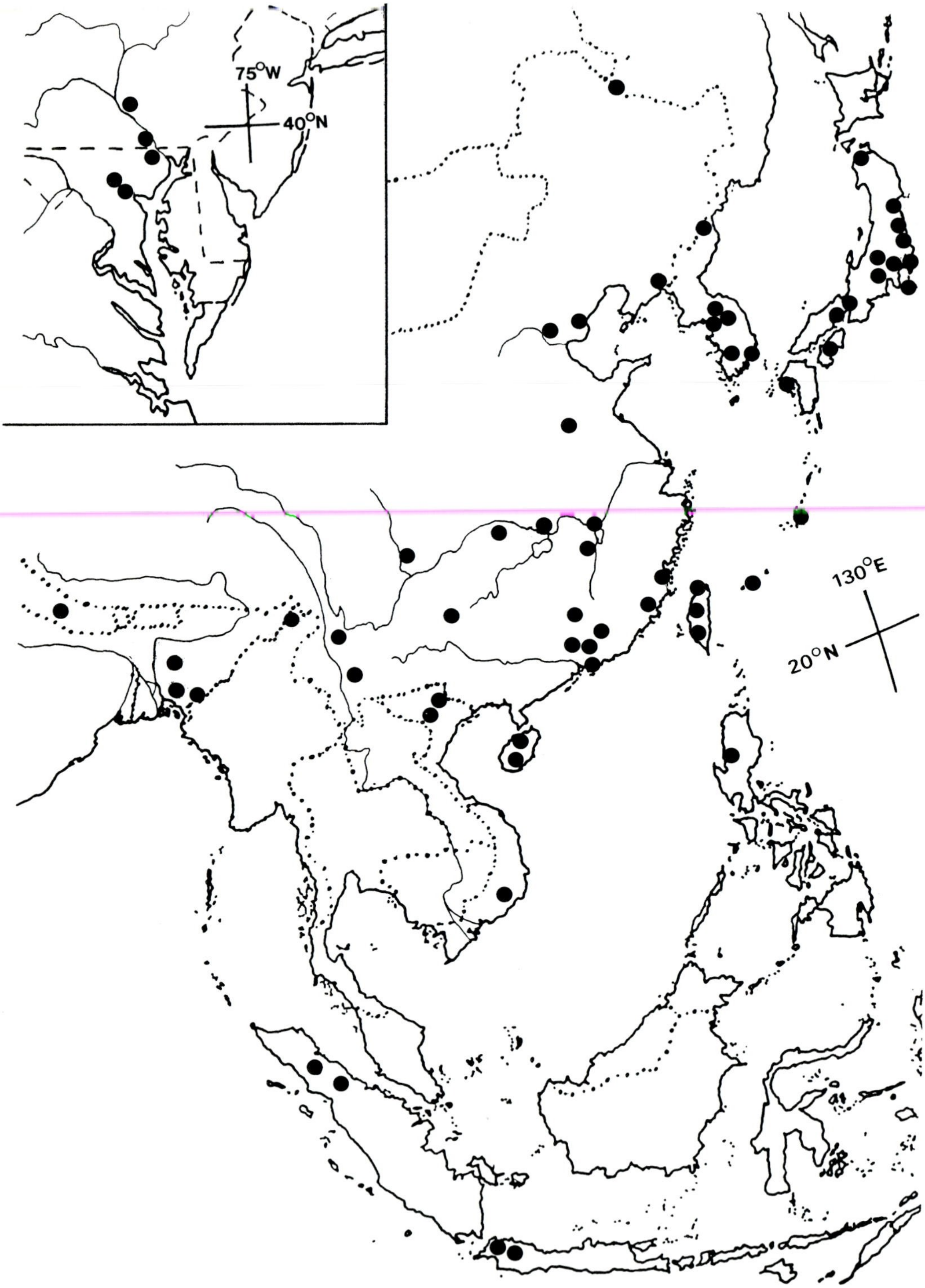

Fig. 13. Distribution of *P. perfoliatum.*

to City, 14 Oct 1960, *Murata 13865* (MAK); Kyoto City, Ukyo-ku, nr. Katsura, sunny place along way, 3 Nov 1965, *Murata 19699* (E, L, MSC, NY). MIYAGI: Shiroishi City, Nishibori, Kouriyama, 40 m, 23 Sep 1979, *Houya s.n.* (TUS 67571); Sendai, Nanakita-gawa, 6 Oct 1981, *Naito s.n.* (TUS); Kami-gun, Nakaniida-machi, Naruse River, 23 Sep 1978, *Naito & Ishikawa s.n.* (TUS); Sendai, Idohama, 2 Sep 1971, *Naito & Ohba 71908* (TUS); Sendai, Nigatake, 17 Sep 1955, *Nakamura s.n.* (TUS 16907). NAGANO: Kitasaku-gun, Asama-yama, 9 Sep 1908, *Makino 16630* (MAK); Sasanoi-cho, Nishiyokota, 31 Aug 1897, *Minemura 456* (MAK). NAGASAKI: Nagasaki, Aug 1863, *Maximowicz s.n.* (LE). OSAKA: Toyanaka-shi, Kumanoda, sunny wayside, 21 Nov 1965, *Togashi 700* (E, L, MSC, NY). RYUKYU: Okinawa Isl., Kunigami-gun, Ginoza-son, Kanna, 50 m, 25 Aug 1983, *Tateishi 9694* (TUS); Yaeyama-gunto, Ishigaki Isl., open waste field just W of village (Kabira), 3 Sep 1951, *Walker & Tawada 7229* (L). SAITAMA: Koshigaya, 28 Sep 1950, *Ohwi 129* (DAO, F, L, MT, NY, P, TUS). TOKYO: Kunitachi-chi, in 1930, *Makino 16623* (MAK); Miyake-zima, Kamizuki-mura, 9 Jul 1938, *Makino 16628* (MAK).

KOREA. CHUNGNAM: Kongju, 28 Jun 1931, *Chung s.n.* (SNU). HAMBUK: Moosan-gun, Samjang-myun, without date, *Makino?* (MAK 16618). KYUNGBUK: Tongrai, 18 Aug 1954, *Col.?* (DUKE). KYUNGGI: Kumkok, in open, wet ground along abandoned rice field, 3 Aug 1981, *Park 205* (BH); Ichon, 8 Aug 1954, *Col.?* (DUKE). SEOUL: Without specific locality, Jul 1886, *Kalinowsky s.n.* (LE); in wet, shady places along roadside nr. Seoul National University Campus, 17 Jul 1981, *Park 201* (BH); Seoul National University Campus, in open, disturbed waste ground, 27 Jul 1981, *Park 202* (BH); Chang-kyung Palace, 21 Jun 1930, *To 1598* (SNU); Mt. Bukhan, 4 Jul 1954, *Col.?* (DUKE); without specific locality, in 1948, *Col.? 9094* (F).

NEPAL. Without specific locality, in 1821, *Wallich s.n.* (NY); Pokhara, 3000 ft, 28 Oct 1954, *Sykes & Williams 8310* (E).

PHILIPPINES. LUZON: Benguet Prov., Baguio, Mar 1904, *Elmer 5793* (NY); Benguet Prov., Sablan, Apr 1904, *Elmer 6261* (NY); Benguet Prov., Baguio, Mar 1907, *Elmer 8631* (NY); without specific locality, Dec 1908, *Jones 40* (F).

TAIWAN. KAOHSIUNG: Shang-ping, 22 Aug 1972, *Moo 2158* (TUS). NANTOU: Chitou, 1100 m, 8 Oct 1966, *Van Steenis 20770* (L). TAINAN: Kuan-tzu-ling, hillside, 19 Oct 1965, *Kao 6308* (L, TAI). TAIPEI: Tansui, in 1864, *Oldham 420* (LE); vic. Taipei, 12 Jul 1929, *Tanaka s.n.* (BM); Tansui, 31 Aug 1930, *Tanaka 70* (NY); Taipei, 30 May 1932, *Tanaka & Shimada 11078* (F, NY).

U.S.A. MARYLAND. Baltimore Co.: Gwynns Falls, stream nr. Liberty Rd. and Baltimore Beltway intersection, 21 Oct 1977, *Latane s.n.* (UNC); Owings Mills, Bonita Ave., wet place nr. stream, 17 Aug 1982, *Park 310* (BH). Harford Co.: The Rocks, along Deer Creek, 22 Sep 1971, *Reed 91891* (UNC). Prince George's Co.: College Park, weed in shrubs at NW side of Health Center at roadside, University of Maryland Campus, 17 Jul 1980, *Hill 9343* (NY); College Park, shady place nr. University of Maryland Campus, 17 Aug 1982, *Park 309* (BH). PENNSYLVANIA. Dauphin Co.: Harrisburg, in open, shady places nr. Harrisburg Area Community College parking lot, 19 Oct 1981, *Park 250* (BH); Harrisburg, shady, disturbed area along East 22, nr. Cameron St., 19 Oct 1981, *Park 251* (BH); Harrisburg, in big, open area along Cameron St., nr. Interstate 81 junction, 19 Oct 1981, *Park 252* (BH); Harrisburg, in shady places nr. Harrisburg Area Community College parking lot, 13 Aug 1982, *Park 301* (BH). York Co.: ½ mi S of Stewartstown, climbing in trees in old orchard, 7 Aug 1947, *Wahl 3495* (FLAS, LAF); Gable's Nursery, S of Stewartstown, 13 Sep 1955, *Williams s.n.* (UNC).

U.S.S.R. AMUR: Blagoveshchensk, 7 Aug 1898, *Karo 156* (E, L, P); without specific locality and date, *Maximowicz s.n.* (L, P.)

VIETNAM. ANNAM: Lang-biang, 1100 m, 25 Oct 1922, *Balansa 3971* (P). TONKIN: Without specific locality, Sep 1885, *Balansa 537* (P); without specific locality, Oct 1885, *Balansa 1021* (P); without specific locality, Nov 1890, *Balansa 4889* (P); Prov. Phuto, 14–19 Apr 1914, *Fleury 32078* (P); without specific locality, Jun 1906, *Mouret 322* (P); Cho Ganh?, Nov 1922, *Pételot s.n.* (P); Cao Bang, without date, *Col.?* (P).

2. **Polygonum senticosum** (Meisner) Franchet & Savat., Enum. pl. Jap. **1:** 410. 1875.

Chylocalyx senticosus Meisner, Ann. Mus. Bot. Lugduno-Batavum **2:** 65. 1865. *Persicaria senticosa* (Meisner) Gross ex Nakai, Fl. Saishu & Kwan Isl. 41. 1914. *Truellum senticosum* (Meisner) Danser, Bull. Jard. Bot. Buitenzorg, sér. 3. **8:** 31. 1926, nom. inval. Type. Japan. "In arenosis litoralibus principatus Fizen ins. Kiusiu," without date, *Pierot s.n.* (lectotype, L!, designated by Park, Brittonia **38:** 396. 1986). Syntypes. Japan. Without specific locality and date, *Siebold, Buerger* (L!, NY!; see Park, Brittonia **38:** 397. 1986).

Truellum japonicum Houttuyn, Natuurlyke historie **8:** 427 (genus description), pl. 48, fig. 1 (species name in index of plates (not paged: 4th page)). 1777. *Polygonum truellum* Houttuyn ex Koidzumi, Bot. Mag. (Tokyo) **40:** 334. 1926, nom. illeg. *Persicaria truellum* Honda *in* Miyabe & Kudo, J. Fac. Agric. Hokkaido Univ. **26:** 515. 1934, nom. illeg. Type. Houttuyn's plate cited above (lectotype, designated by Park, Brittonia **38:** 396. 1986).

Polygonum babingtonii Hance, Ann. Sci. Nat. Bot., sér. 5. **5:** 239. 1866. Type. China. "In provincia Cantoniensi," without date, *Sampson 7497* (lectotype, K!, designated by Park, Brittonia **38:** 396. 1986). Syntype. Japan. *Babington s.n.* (not seen).

Polygonum typhoniifolium Hance, Ann. Sci. Nat. Bot., sér. 5. **5:** 239. 1866. Type. China. "E. provincia Fokien," in 1857, *Gregory 1269* (not seen; see Forbes & Hemsley, J. Linn. Soc., Bot. **26:** 349. 1891).

Scandent or climbing *herb,* 1–2.5 m tall; stem much branched, reddish-purple at maturity, often finely pubescent with minute simple unicellular hairs in upper portion, angled, densely armed with recurved prickles along the angles; prickles stout, broad-based, 1–1.5 mm long. *Leaves* long-petioled; blades thin, membranous, light green to green, triangular, 4.5–8.5 cm long, 4.5–8.8 cm wide, subacute to acuminate at the apex, at the base broadly cordate to weakly hastate or deeply sagittate with subacute to acuminate triangular lobes; margins and midveins beneath retrorsely prickly; both surfaces glabrous to densely pubescent with fine simple unicellular hairs to 0.1 mm long; petioles 3.3–7.5 cm long, retrorsely prickly, often reddish-purple at maturity. *Ocreae* foliaceous, green, somewhat hyaline at the base, 0.5–1 cm long, almost surrounding the node. *Inflorescences* terminal or in the axils of upper leaves, elongate, 4.5–18 cm long, often branched 1–2 times, retrorsely prickly at least in lower portion, glabrous to densely pubescent with glandular hairs; glandular hairs stalked, ca. 0.5 mm long; inflorescence bracts resembling ocreae but much reduced, 2–3 mm long, often associated with reduced leaves; each inflorescence branch terminated by a dense headlike cluster of 3–15 flower fascicles. *Flower fascicles* 2- to 3-flowered, closely spaced; bracts rhomboid, ciliate, ca. 2 mm long; bracteoles broadly elliptic, entire, glabrous, 2–3 mm long; pedicels very short, 1.5–2.5 mm long, glabrous, included in bracteoles; perianth deeply 5-parted, pink to red, often whitish-pink at the base, 3–5 mm long; tepals elliptic, not becoming fleshy; stamens eight, inserted at the base of perianth in two whorls; inner stamens three, ca. 1.5 mm long; outer stamens five, ca. 1 mm long; style one, 3-cleft up to middle, ca. 2 mm long, included; stigmas three. *Achenes* weakly trigonous, dark brown to black, dull, rugose, 2–3.5 mm long.

In central China, plants with auriculate foliaceous ocreae occur sporadically and have been formerly recognized as a distinct species (Léveillé, 1912; Léveillé & Vaniot, 1902); these plants, however, are not consistently distinguishable by any characters other than ocrea shape (Park, 1986a). In the view of the overall floral and vegetative similarities to *P. senticosum,* it is recognized as a taxonomic variety. The two varieties may be distinguished as follows:

1. Plants pubescent with simple unicellular hairs; ocreae reniform, ciliate. var. *senticosum.*
1. Plants glabrous; ocreae auriculate, eciliate. var. *sagittifolium.*

2a. P. senticosum (Meisner) Franchet & Savat. var. **senticosum.**

Stem usually sparsely to moderately pubescent with simple unicellular hairs in upper portion. *Leaves* triangular, 4.5–7.8 cm long, 4.5–8 cm wide, broadly cordate to weakly hastate with subacute triangular basal lobes, sparsely to densely pubescent with fine simple unicellular hairs on both surfaces, rarely glabrous; petioles often bearing simple unicellular hairs in upper portion; ocreae reniform, ciliate at the margin. *Inflorescences* usually branched 1–2 times, sparsely to densely glandular-pubescent in upper portion.

Distribution. In open sun or partial shade; occurring naturally in streambank thickets; common as a weed in rice paddies, abandoned fields, roadsides, and other disturbed sites; common in Japan, Korea, Manchuria, and China (Fig. 14); known from two localities in the United States (Portland, Oregon).

Additional specimens examined. CHINA. ANHWEI: Chien Shan Heisen, Tien Chu Shan, under bamboo forest, 250 m, 16 Jul 1936, *Fan & Li 116* (GH, L). CHEKIANG: Without specific locality, 2 Jul 1953, *Lee 522* (PE). FUKIEN: Foochow and vic., 11 Sep 1926, *Chang 4940* (MO); Foochow City, Wooshih-shan, in wet places, 26 Aug 1923, *Chung 2362* (GH); Yenping, Huang-chin-shan, roadside, 130 m, 19 Aug 1924, *Chung 3023* (GH); without specific locality, Apr–Jun 1905, *Dunn 3456* (GH); Hinghwa and vic., 4 Sep 1926, *Pi & Mo 6459* (MT). KIANGSI: Hsing Tsu Hsien, Lu-shan, 25 Aug 1932, *Cheo 198* (E); Kiennan Dist., Sai Hang Cheung nr. Tung Lei Village, 1–29 Aug 1934, *Lau 4165* (GH); Lungnan Dist., Oo Chi San, nr. Lam Uk Village, 1–25 Oct 1934, *Lau 4800* (GH). KWANGSI: Without specific locality and date, *Wang 41088* (A). KWANGTUNG: Canton and vic., 13 Aug 1917, *Levine 1363* (GH, MO); Yaoshan, North River, on the hill between Kuei-tow and Tsau-tou-ping, 300 m, 26 Apr 1930, *Sin s.n.* (NY); Lung Tau Shan, in village and Ha-tso-kok Bay, 13 Jun, 10 Jul 1924, *To & Tsang 12466 & 12874* (MO, NY). LIAONING: Without specific locality, 15 Aug 1909, *Yabe s.n.* (NSM). SHANTUNG: Niu Shu Tai, Lao Shan, streamside, 4 Jul 1930, *Fan & Li 2703* (E, GH, MT, NY).

JAPAN. AOMORI: Prov. Mutsu, Bushidomari in Shimokita Peninsula, 10 Aug 1964, *Yoshioka & Kaneko s.n.* (TUS 16945). FUKUI: Obama-shi, Wakasa, 20 Aug 1910, *Izaki s.n.* (MAK 17239). FUKUSHIMA: Futaba-gun, Hirono-machi, Boda, 30 m, 7 Aug 1980, *Nemoto 583* (TUS). GIFU: Prov. Mino, 13 Jul 1928, *Seki 4574* (GH); Prov. Mino, 10 Sep 1921, *Shiota 3877* (GH); 24 Jul 1930, *3818* (GH); Prov. Mino, Aug 1933, *Mizuno 8216* (GH). HIROSHIMA: Hiroshima, Kusatsu, 21 Jun 1959, *Ando s.n.* (MT). HOKKAIDO: Jesso Island, nr. Hakodate, in 1861, *Albrecht s.n.* (LE). HYOGO: Harima, 23 Jul 1903, *Arimoto s.n.* (GH); Yamamoto in

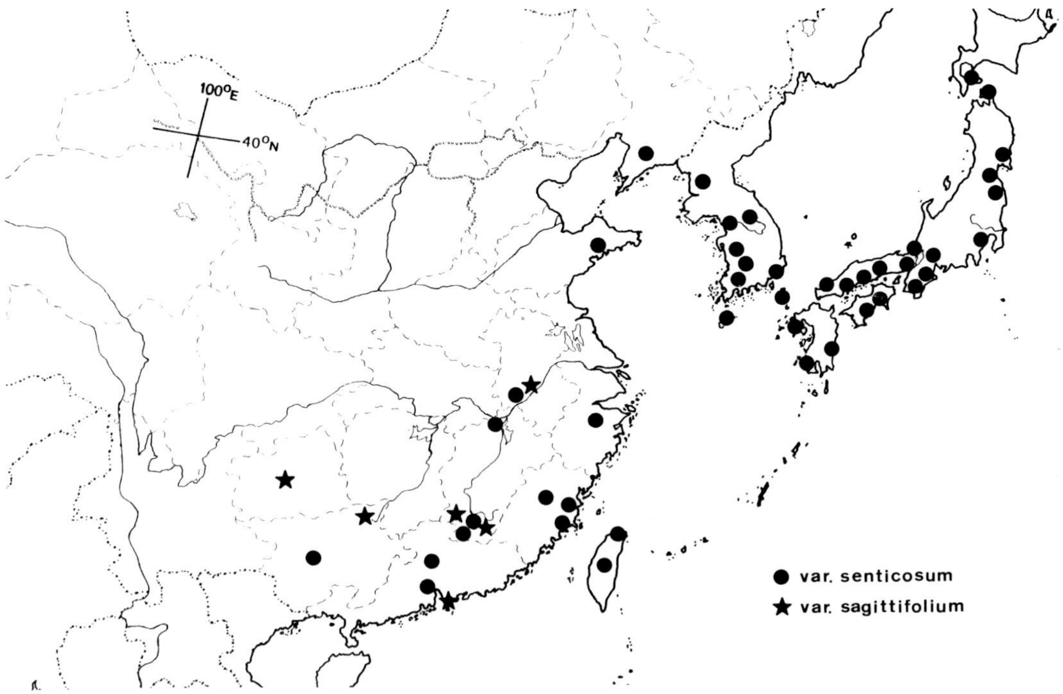

Fig. 14. Distribution of *P. senticosum.*

Settsu, 2 Sep 1953, *Togashi 807* (BM, GH, MAK, MT, NY, TUS). **Iwate:** Kesennuma-shi, 27 Sep 1900, *Makino 17228* (MAK). **Kagoshima:** Prov. Osumi, Amboo, Yaku-machi Is., Yaku-shima, 13 May 1962, *Furuse s.n.* (A); Prov. Satuma, in Siroyama, 19 May 1920, *Kimura s.n.* (TUS 16939); Yokohama City, without date, *Lavatov? s.n.* (A); Nase-shi, 4 Apr 1901, *Makino 17248* (MAK); Yokohama City, Kawashima-mura, 6 Jun 1901, *Makino 17233* (MAK); Yokohama City, Kohoku-ku, 4 Sep 1912, *Makino 17253* (E); Yokohama City, in 1862, *Maximowicz s.n.* (GH, LE, P); "circa Yokohama," without date, *Savatier 1044* (P); Miura-shi, Tsurugi-zaki, 8 Oct 1981, *Takahashi 1942* (TUS). **Kochi:** Kami-gun, Monobe-mura, en route from Sejrou to Okanouchi, 350–500 m, 21 May 1974, *Kurosaki 6264* (TUS); Aki-shi, 3 Jun 1892, *Makino 17241* (MAK); Kochi-shi, Katsura-hama, 17 Oct 1970, *Shoma et al. s.n.* (MO, TUS). **Kyoto:** Soraku-gun, Wazuka-cho, Mt. Jyubu, in deciduous forest, 300–600 m, 17 Sep 1978, *Tsugaru 5106* (MO). **Mie:** Ise-shi, Futami, Jun 1893, *Makino 17237* (MAK). **Miyagi:** Rikuzen, Isawa-gun, Mizusawa, 29 Jul 1931, *Iwabuchi s.n.* (TUS); Rikuzen, Miyatojima, 22 Sep 1957, *Kyogoku s.n.* (TUS 16934); Sendai, Kunimi-toge, ca. 200 m, 5 Aug 1982, *Ohashi 8418* (TUS); Rikuzen, Tsubakijima, 25 Aug 1955, *Sugaya et al. s.n.* (TUS 16944). **Miyazaki:** Miyazaki-shi, May 1902, *Yamana s.n.* (MAK 17246). **Nagasaki:** Nagasaki, Kumamoto, 10–22 May 1863, *Maximowicz s.n.* (L); Nagasaki, in 1862, *Oldham 695* (GH, L); Tsushima Island, in 1859, *Wilford s.n.* (GH). **Nara:** Nara-shi, Minakami-ike Pond, E of Saidaiji, 28 Sep 1969, *Seto 18514* (NLU). **Okayama:** Mimasaka, 4 Aug 1903, *Arimoto s.n.* (MO). **Osumi:** Sata-misaki, 24 May 1960, *Sugaya & Sohma s.n.* (TUS 16940). **Ryukyu:** Okinawa Isl., 3 mi N of Nago, sandy soil of upper beach and along roadside, 23 Jul 1951, *Walker et al. 6942* (A, L); without specific locality, 1853–1856, *Wright 252* (GH). **Saga:** Kashima-shi, Kashima-mura, 30 Jul 1910, *Omori s.n.* (MAK 17244). **Satsuma:** Kagoshima-gun, Ishiki-mura, Oyamada, 15 Aug 1910, *Morimoto s.n.* (MAK 17249). **Wakayama:** Higashimuro-gun, Kazachi, 21 Jul 1952, *Uno s.n.* (A). **Yamaguchi:** Prov. Nagato, Abe-gun, Nago, 8 Oct 1967, *Sohma 1082* (TUS).

KOREA. "Korea septentrionalis," 16 Aug 1897, *Komarov 558* (NY); without specific locality, Jul 1883, *Perry s.n.* (GH). **Cheju:** Without specific locality, Jun 1911, *Taquet 909?* (E). **Chungbuk:** Mt. Sokni, nr. Beopju Temple, 25 Aug 1969, *Migo s.n.* (TUS 45549). **Chungnam:** Kong-ju, 20 Jul 1931, *Chung s.n.* (SNU). **Kangwon:** Yong-pyung, Mt. Balwang, wet place along roadside, 21 Aug 1981, *Park 450* (BH). **Kyunggi:** Kwang-reung, 16 Jun 1944, *Chung s.n.* (SNU); without specific locality, 18–20 Aug 1897, *Komarov s.n.* (LE); Mt. Chun-ma, wet place along roadside, 20 Jul 1981, *Park 411* (BH); Kum-kok, open, disturbed place nr. abandoned field, 20 Jul 1981, *Park 412* (BH); Kwang-reung, wet place nr. small stream, 22 Jul 1981, *Park 420* (BH). **Pusan:** Without specific locality, 15 Aug 1947, *To s.n.* (SNU 1621). **Pyungyang:** Without specific locality, 30 Aug 1937, *Smith s.n.* (GH). **Seoul:** Chang-kyung Palace, 12 Aug 1943, *Chung s.n.* (SNU).

TAIWAN. "Bankinsing," Feb 1914, *Faurie 764* (P); "Mt. Kwannin," without date, *Faurie 1184* (P); without specific locality, in 1915, *Faurie 8116* (GH); without specific locality, in 1930, *Makino?* (MAK 17250); without specific locality, in 1864, *Oldham s.n.* (GH, P). HUALIEN: "Hwalien," 20 Aug 1929, *Suzuki 1698* (MO). NANTOU: Hoshi tract, 29–31 Mar 1961, *Huang 2209* (MSC). TAIPEI: Mt. Kuan-ying, 12 Sep 1965, *Chuang 3914* (MO); Mientienshan, 6 Jul 1974, *Kuo 5408* (TAI); Taipei, 10 Jul 1927, *Sasaki 21576* (NY). U.S.A. OREGON. Multnomah Co.: Portland, Albina, 30 Nov 1904, *Suksdorf 3284* (WS); Portland, Linnton, on ballast, 28 Aug 1910, *Suksdorf 1607* (WS).

2b. Polygonum senticosum var. **sagittifolium** (Léveillé & Vaniot) Park, Brittonia **38:** 218. 1986.

Polygonum sagittifolium Léveillé & Vaniot, Bull. Acad. Int. Géogr. Bot. **11:** 342. 1902. Type. China. Kweichow: "Environs de Tsin-gay à Tchao-se, dans les rocailles," 7 Sep 1899, *Laborde 2720* (holotype, E!). *Polygonum darrissii* Léveillé, Repert. Spec. Nov. Regni Veg. **11:** 297. 1912. *Truellum darrissii* (Léveillé) Sojak, Preslia **46:** 145. 1974. Type. China. "Dans les fourrés à Hong-Kouang," Jul 1911, *Esquirol 2866* (holotype, E!).

Plant completely glabrous except for recurved prickles. *Leaves* narrowly triangular, 5–8.5 cm long, 3–5 cm wide, acuminate at the apex, weakly hastate to deeply sagittate with acuminate triangular lobes at the base. *Ocreae* auriculate, eciliate; auricles linear lanceolate to lanceolate, 0.5–1 cm long. *Inflorescences* usually unbranched.

Distribution. Known only from China (Anhwei, Hunan, Kweicho, and Kwangsi; Fig. 14).

Additional specimens examined. CHINA. ANHWEI: Wang-shan, in 1973?, *Chow 72* (MO). HUNAN: Yang-shan, 700 m, 12 Jul 1935, *Fan & Li 240* (GH); "in monte Yün-schan prope urbem Wukang ad margines silvarum," 1250 m, 9 Aug 1918, *Handel-Mazzetti 2599* (E, P). KWANGSI: Without specific locality and date, *Wang 41088* (A); northern Kwangsi, Ling-chuan Dist., Sam Kok Ku, Ta-ling, Yang-wu Village and vic., 21–30 Jul 1937, *Tsang 27926* (A, US).

3. Polygonum debile Meisner, Ann. Mus. Bot. Lugduno-Batavum **2:** 63. 1865, excl. var. *hastatum*. *Persicaria debilis* var. *triangularis* (Meisner) Gross ex Nakai, Fl. Saishu & Kwan Isl. 40. 1914. *Persicaria triangularis* (Meisner) Nakai, Rigakkai **24:** 300. 1926. *Truellum debile* (Meisner) Sojak, Preslia **46:** 145. 1974. Type. Japan. Without specific locality and date, *Siebold s.n.* (lectotype, L!, designated by Park, Brittonia **38:** 397. 1986; isolectotype, NY!).

Polygonum thunbergii var. *coreana* Léveillé, Repert. Spec. Nov. Regni Veg. **8:** 171. 1910. Type. Korea. Cheju: Cheju Isl., "in silvis Piento-Hoatien," 19 Sep 1908, *Taquet 1288* (lectotype, E!, designated by Park, Brittonia **38:** 397. 1986; isolectotype, L!). Syntypes. Korea. Cheju: Cheju Isl., "in sylvis Yengsil," 17 Aug 1908, *Taquet 1329* (E!); Cheju Isl., *Taquet 1319 & 1324* (E; see Lauener, Notes Roy. Bot. Gard. Edinburg **40:** 197. 1982).

Ascending to erect *herb* arising from fibrous root system, 30–50 cm tall; stem light green, slender, branched, somewhat decumbent at the base, angled, glabrous, almost unarmed, often bearing a few recurved prickles ca. 0.5 mm long along the angles; lower nodes often producing roots. *Leaves* long-petioled; blades thin, membranous, triangular, 1.5–6 cm long, 1.5–6.8 cm wide, acuminate at the apex, truncate to weakly hastate with acuminate triangular lobes at the base, antrorsely ciliate on upper margin, retrorsely ciliate on lower margin; upper surfaces green, bearing a few appressed bristlelike simple multiseriate hairs to 0.5 mm long, often with a few stellate hairs; stellate hairs sessile, 3- to 5-rayed; lower surfaces light green, sparsely to moderately pubescent with stellate hairs identical to those of upper surfaces; midveins and major veins beneath antrorsely prickly; prickles 0.5–1 mm long; petioles 0.5–3.5 cm long, almost glabrous. *Ocreae* short, 2–5 mm long, somewhat funnelform, scarious but usually with a small dilated foliaceous limb, often bearing short minute recurved prickles at the base. *Inflorescences* terminal or in the axils of upper leaves, commonly 2–3 inflorescences arising from an axil, elongate, slender, often branched, 2.3–3 cm long, often glandular-pubescent; glandular hairs stalked, ca. 0.5 mm long; inflorescence bracts resembling ocreae but frequently without foliaceous limbs and reduced, often ciliate at the margin; each inflorescence branch terminated by a loose, somewhat interrupted panicle of 2–3 flower fascicles. *Flower fascicles* 2- to 3-flowered, widely spaced; bracts broadly elliptic to ovate, 1.5–2 mm long, acuminate at the apex, usually glabrous, ciliate at the margin; bracteoles lanceolate, 2.5–3 mm long, partly exserted, glabrous; pedicels short, 1–1.5 mm long, glabrous, included in bracteoles; perianth 5-parted, white, 3–4 mm long; tepals elliptic, not becoming fleshy; stamens commonly 7 (5–8), inserted at the base of perianth in two whorls; inner stamens three, ca. 1.5 mm long; outer stamens 2–5, ca. 1 mm long; style one,

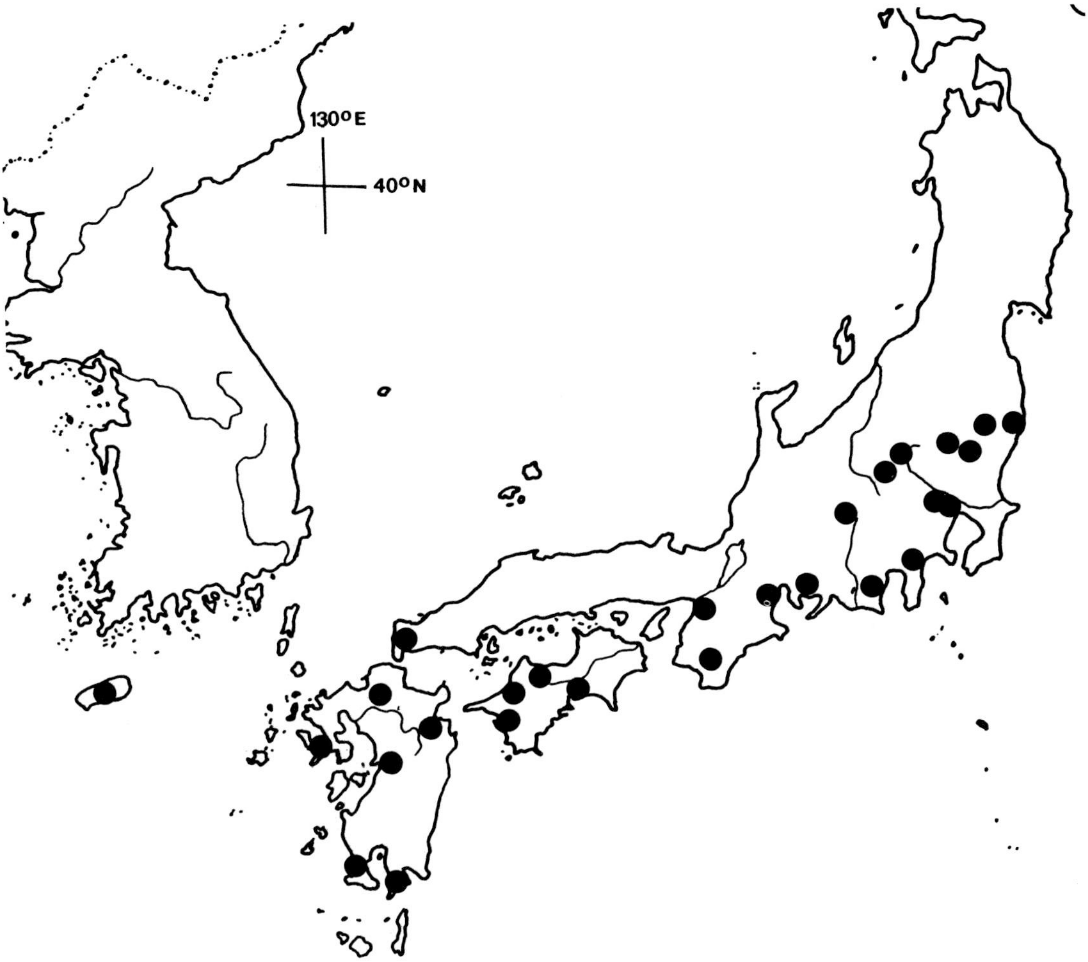

Fig. 15. Distribution of *P. debile*.

3-cleft up to middle, ca. 1 mm long, included; stigmas three. *Achenes* trigonous, ovoid in outline, brown, acute at the apex, smooth, shiny, ca. 3 mm long.

Distribution. Occurring naturally in damp places in forests; known only from central and southern Japan, and southern Korea (Fig. 15).

Polygonum debile is apparently related to *P. senticosum,* as indicated by its broadly triangular leaves with truncate to weakly hastate bases and ocreae with foliaceous limbs. Differences include ascending to erect habit, loose interrupted paniclelike inflorescences, and smooth shiny achenes. In addition, leaves of *P. debile* are stellate-pubescent, whereas those of *P. senticosum* are either glabrous (var. *sagittifolium*) or pubescent with simple unicellular hairs (var. *senticosum*). The presence of stellate hairs, frequent loss of foliaceous limb on ocrea, and ascending to erect habit found in *P. debile* also suggest a possible relationship with *P. thunbergii,* but it is morphologically closer to *P. senticosum.*

As Steward (1930) pointed out, the type specimen of var. *hastatum* Meisner (*Hook. f. & Thomson 58*) appears to be the poorly developed form of *P. thunbergii* var. *thunbergii.*

Additional specimens examined. JAPAN. Without specific locality and date, *Bunge? s.n.* (L). AICHI: Prov. Mikawa, Kitashitara-gun, Toyone-mura, Mt. Causu, 21 Aug 1966, *Kato s.n.* (TUS). EHIME: Iyo, Kuma-cho, Iwaya-ji, 21 Oct 1970, *Kimura N153.34.3* (TUS); Onsen-gun, Shigenobu-cho, 24 Jul 1897, *Makino 13399* (MAK); Uwazima-shi, Nametoko-yama, 5 Nov 1907, *Yamada s.n.* (MAK 13445). FUKUOKA: Tagawa-gun, Hiko-zan, 30 Aug 1906, *Makino 13449* (DAO, MAK).

FUKUSHIMA: Nishishirakawa-gun, Oct 1909, *Imai s.n.* (MAK 13392); Iwaki-shi, Ogawa-machi, Ushigogawa, 30 Jul 1983, *Ohashi 8535* (TUS). GUNMA: Numata, Akagi-yama, 10 Sep 1911, *Honda s.n.* (MAK 13394); Tano-gun, Ueno-mura, Fujidoo, 750–900 m, 30 Aug 1978, *Murata et al. 6258* (TUS); Ikaho in Kotsuke, 4 Sep 1950, *Ohwi 76* (DAO, F, L, MT, MICH, NY, P, TUS, US). HYOGO: Prov. Harima, 30 Sep 1951, *Yamazaki 2958* (MT). KAGOSHIMA: Kagoshima-gun, Sakurazima, in 1913, *Tashiro s.n.* (MAK 13448). KANAGAWA: Odawara, Hakone, 25 Sep 1886, *Makino 13440* (MAK); Hakone, 30 Aug 1871, *Savatier s.n.* (P). KOCHI: Agawa-gun, Agawa-mura, Okunanokawa, Nigootsutoge, 30 Jul 1959, *Yamanaka 4721* (TUS); Tosa, Nanokawa, 18 Jul 1892, *Col.?* (US). KUMAMOTO: Kikuchi-gun, Hukabayama, 23 Sep 1906, *Kozuma s.n.* (MAK 13401). MIE: Mie University Hirakura Exp. Forest, Okumura, Yahata-mura, Isshi-gun, 500–600 m, 18 Sep 1962, *Kanai 6776* (MT); Suzuka-shi, 23 Sep 1906, *Kawai s.n.* (MAK 13398). NAGANO: Togakushichuja, 20 Aug 1960, *Koyama s.n.* (DAO). NAGASAKI: Nagasaki, 1 Sep 1863, *Maximowicz s.n.* (BM, K, L, LE, NY, P). OITA: Kujusan, Jusanmagari, Yobikogawa, 30 Sep 1961, *Togashi 7351* (F). OSAKA: Prov. Kawachi, Mt. Kongosan, en route from Chihaya to Fushimitoge, 650 m, 23 Jul 1962, *Murata 27061* (US); Mt. Kongosan in Kawachi, 17 Sep 1953, *Togashi 883* (DAO, E, K, LE, MICH, MT, NY, TUS, US). SAITAMA: Musashi-koku, Takao-san, 2 Sep 1914, *Endo s.n.* (TUS). SHIGA: Mt. Hiei, nr. Komponchudo Temple, ca. 750 m, 5 Jul 1964, *Kasapligil 3635* (K); Mt. Hiei, wet shady place in woods, 19 Aug 1963, *Murata 18400* (E, L, LE, MSC, NY). SHIZUOKA: Fujiyama, 22 Aug 1890, *Faurie 6659* (K); en route from the top of Mt. Fudou to the foot of Mt. Morosawa, Hon-kawanecho, Haibara-gun, on humus slope in deciduous forest, 1300 m, 30 Jul 1978, *Koyama et al. 101* (BM); Abegun, Umbashima-mura, en route from Umegashimaonsen to Abe-toge, 8 Aug 1967, *Wakabayashi 97* (MSC). TOCHIGI: Prov. Shimotsuke, Nikko, Hosoo, 28 Jul 1891, *Hori 13466* (DAO); Kurobane-machi, W side of Mt. Hanakame-yama, ca. 400 m, 7 Oct 1983, *Nagayama 424* (TUS); Nikko, Mimotsuke, 17 Aug 1910, *Col.?* (US). TOKYO: Musashino, Nishitamagawa-gun, Komazawa, 15 Sep 1957, *Mizushima s.n.* (MAK 1865). WAKAYAMA: Prov. Kii, Okumotori-yama, between Zizochaya, Higashimuro-gun, 16 Oct 1960, *Koyama s.n.* (MSC). YAMAGUCHI: Kuka-gun, Nishiki-machi, Nishiki-dake, 27 Sep 1980, *Kunio 45817* (TUS).

KOREA. CHEJU: Cheju Isl., "in sylvis," 800 m, Aug 1911, *Taquet 9904* (TI).

4. Polygonum thunbergii Siebold & Zucc., Abh. Math.-Phys. Cl. Königl. Bayer. Akad. Wiss. **4(3):** 208. 1846.

Polygonum arifolium sensu Thunberg, Fl. jap. 168. 1784; non L., 1753. *Helxine arifolia* Raf., Fl. tellur. **3:** 94. 1837. *Tracaulon thunbergii* (Siebold & Zucc.) Greene, Leafl. Bot. Observ. Crit. **1:** 22. 1904. *Persicaria thunbergii* (Siebold & Zucc.) Gross ex Nakai, Fl. Saishu & Kwan Isl. 41. 1914. *Truellum thunbergii* (Siebold & Zucc.) Sojak, Preslia **46:** 149. 1974. Type. Japan. Nagasaki: "Circum Nagasaki vulgare," without date, *Thunberg s.n.* [holotype, UPS (microfiche at BH!); see Park, Brittonia **38:** 398. 1986].

Polygonum hastatotrilobum Meisner, Ann. Mus. Bot. Lugduno-Batavum **2:** 62. 1865. *Polygonum thunbergii* var. *hastatotrilobum* (Meisner) Maxim. ex Franchet & Savat., Enum. Pl. Jap. **2:** 475. 1878. *Tracaulon hastatotrilobum* (Meisner) Greene, Leafl. Bot. Observ. Crit. **1:** 22. 1904. Type. Japan. "In scrobibus viarum ins. Nippon et Kiusiu," without date, *Pierot 263* (lectotype, L!, designated by Park, Brittonia **38:** 398. 1986; isolectotype, L!). Syntypes. Japan. *Siebold, Buerger* (not seen).

Polygonum debile var. *hastatum* Meisner, Ann. Mus. Bot. Lugduno-Batavum **2:** 63. 1865. Type. India. Assam: Mt. Khasia, "in region. temper.," 6–7000 ft, *Hook. f. & Thomson 58* (lectotype, NY!, designated by Park, Brittonia **38:** 398. 1986; isolectotype, GH!).

Polygonum stoloniferum Schmidt, Mém. Acad. Imp. Sci. Saint Pétersbourg, sér. 7. **12(2):** 168. 1868. *Polygonum thunbergii* var. *stoloniferum* (Schmidt) Makino, Bot. Mag. (Tokyo) **24:** 74. 1910. *Persicaria thunbergii* var. *stolonifera* (Schmidt) Gross ex Nakai, Fl. Saishu & Kwan Isl. 41. 1914. Type. U.S.S.R. Sakhalin: "In Moossümpfen, Traiziska Anf.," Aug 1860, *Schmidt s.n.* (lectotype, NY!, designated by Park, Brittonia **38:** 398. 1986; isolectotype, GH!). Syntypes. U.S.S.R. Sakhalin: Tunaitscha, 3 Sep 1861, *Glehn s.n.* (GH!); Truotoga, 4 Aug 1861, *Glehn s.n.* (not seen).

Polygonum pteropus Hance, J. Bot. **7:** 167. 1869. Type. China. "Nankau dictas, secus viam a Peking in Mongoliam ducentem," *Williams 14890* (not seen; see Steward, Contr. Gray Herb. **5(88):** 84. 1930).

Polygonum thunbergii var. *oreophilum* Makino, Bot. Mag. (Tokyo) **27:** 251. 1913. *Persicaria thunbergii* var. *oreophila* (Makino) Nemoto, Fl. Jap. Suppl. 176. 1936. *Persicaria oreophila* (Makino) Hiyama, J. Jap. Bot. **34:** 340. 1959. *Polygonum oreophilum* (Makino) Ohwi, Fl. Jap. 409. 1965. *Truellum oreophilum* (Makino) Sojak, Preslia **46:** 148. 1974. Type. Japan. Prov. Musashi, Mt. Takao, 5 Oct 1913, *Makino s.n.* (not seen; inquiry to MAK and other major herbaria in Japan failed to locate the type).

Scandent to ascending *herb,* 0.3–1 m tall; stem branched, decumbent at the base, angled, often with minute 6- to 8-rayed sessile stellate hairs, sparsely to densely armed with stout recurved prickles to 1.5 mm long; lower nodes often producing stolons. *Leaves* long-petioled; blades broadly triangular to ovate in outline, hastate with two somewhat divergent basal lobes, 3.5–9.5 cm long, 1–4.5 cm wide (middle lobe), subobtuse to acuminate at the apex, ciliate at the margin; both surfaces almost glabrous to moderately pubescent with appressed bristlelike simple multiseriate hairs to 1.5 mm long, often together with varying densities of stellate hairs; stellate hairs of the leaf sessile, 4- to 7-rayed, similar to those of the stem but rays slender and much longer; midveins beneath usually antrorsely prickly; petioles 1.5–5.5 cm long, retrorsely prickly, often with conspicuous foliaceous wings. *Ocreae* short, 0.5–1 cm long, scarious at the base but usually with a foliaceous dilated limb. *Inflorescences* terminal or in the axils of upper leaves, usually elongate, 2–8 cm long, often branched 1–2 times, sparsely to moderately glandular-pubescent in upper portion; glandular hairs stalked, ca. 0.5 mm long; inflorescence bracts resembling ocreae but much reduced; each inflorescence branch usually terminated by a dense headlike cluster of seven to many flower fascicles. *Flower fascicles* 1- to 2-flowered; bracts lanceolate to elliptic, acuminate, long-ciliate at the margin, 2.5–4 mm long; bracteoles elliptic, 3–4 mm long, ciliate at the tip; pedicels short, 2–3 mm long, glabrous, included in bracteoles; perianth 5-parted, reddish or whitish with purple tinge at the tip, rarely pale green, 3–5 mm long; tepals elliptic, obtuse, not becoming fleshy; stamens eight, rarely seven, inserted at the base of perianth in two whorls; inner stamens three, 1–2 mm long; outer stamens 4–5, 1.5–3 mm long; style one, deeply 3-cleft, 1.5–3.5 mm long, often slightly exserted; stigmas three. *Achenes* sharply trigonous, ovoid in outline, yellowish-brown to black, rarely grayish-brown, punctuate or smooth, 3–4 mm long.

I am unable to locate the type specimens of *P. stellatotomentosum* Smith & Ramaswami (Smith et al., 1913), which apparently is synonymous with *P. thunbergii*. From the species description, *P. stellatotomentosum* seems to be a stellate-pubescent form of *P. thunbergii* var. *thunbergii*, which occurs sporadically throughout the range of its distribution.

Polygonum thunbergii is a widespread and variable species. Two taxonomic varieties recognized here are distinguished by their shapes of ocreae and leaves.

1. Leaves truncate to weakly hastate at the base; middle leaf lobe broadly elliptic to ovate; ocrea margin entire to weakly crenate. var. *thunbergii.*
1. Leaves strongly hastate at the base; middle leaf lobe narrower, lanceolate to narrowly elliptic; ocrea margin strongly dentate. var. *maackianum.*

4a. P. thunbergii Siebold & Zucc. var. thunbergii.

Stem glabrous to sparsely stellate-pubescent; lower nodes often producing flowering and fruiting stolons. *Leaves* large, 3.5–9.5 cm long, 1.7–4.5 cm wide (middle lobe); middle lobe broadly elliptic to ovate, subobtuse to acuminate at the apex, often contracted near the base; basal lobes smaller than the middle lobe, triangular to ovate, obtuse to acute; both surfaces almost glabrous to moderately pubescent with appressed bristlelike simple multiseriate hairs, often also bearing a few stellate hairs but rarely densely stellate-pubescent; petioles 2–5.5 cm long, often conspicuously winged. *Ocreae* short, 0.4–1 cm long, scarious but usually with a small foliaceous limb; limbs entire to weakly crenate but not conspicuously dentate, ciliate. *Inflorescence branches* sparsely glandular-pubescent in upper portion, often stellate-pubescent below; bracts and bracteoles usually glabrous. *Achenes* yellowish-brown to brown, rarely grayish-brown, punctuate, somewhat dull, rarely shiny.

Distribution. Usually in partial shade; occurring naturally in wet places along streams; common as a weed along ditches in rice paddies; from Japan and Korea, through eastern Siberia, China, and northern Indochina, to northern India. Common in Japan, Korea, and China (Fig. 16).

Polygonum thunbergii var. *thunbergii* shows considerable variation in leaf shape; from broadly triangular to broadly ovate with weakly truncate to deeply hastate bases. Achene surfaces are also variable, from punctuate and dull to completely smooth and shiny. However, no absolute morphological discontinuities exist in any of these characters, and entire populations are treated as a single variety.

Additional specimens examined. CHINA. ANHWEI: Without specific locality, Jan 1959, *Anhwei Expd. Team 1030* (NSM); Wu Yuan, 16 Aug 1925, *Ching 8849* (US). CHEKIANG: Without specific locality, 27 Sep 1957, *Chang 1498* (NSM); nr. Ningpo, without date, *Oldham 69* (K). HOPEH: Chihli Prov., Changli, 26 Jun 1913, *Clemens 8052* (E). HUNAN: Ma-Ling-Tung, Sinning Hsien, 650 m, 23 Sep 1935, *Fan & Li 536* (BM, GH, L). HUPEH: *Metasequoia* region of Lichuan Xian, vic. Lojiaba on the W side of the valley, ca. 1500 m, 6 Oct 1980, *Bartholomew et al. s.n.* (NY); Shenlungkai, in valley, 1280 m, in 1976, *Chow 76030* (E, UNC); Ichang, without date, *Henry s.n.* (K); Patung Dist., in 1887, *Henry 2848* (K); without specific locality, 1885–1888, *Henry 4883* (US); without specific locality and date, *Henry 5080* (GH); without specific locality and date, *Henry 7414* (GH, K); Ou-fan-Tcian?, 2090 m, Aug 1907, *Si s.n.* (P); west Hupeh, without date, *Wilson 1039* (E, NY). KIANGSI: Lungnan Dist., Oo Chi Shan nr. Lam Uk Village, 1–25 Oct 1934, *Lau 4859* (GH, US). KIRIN: Kirin, meadow, 20 Aug 1931, *Chen 377*

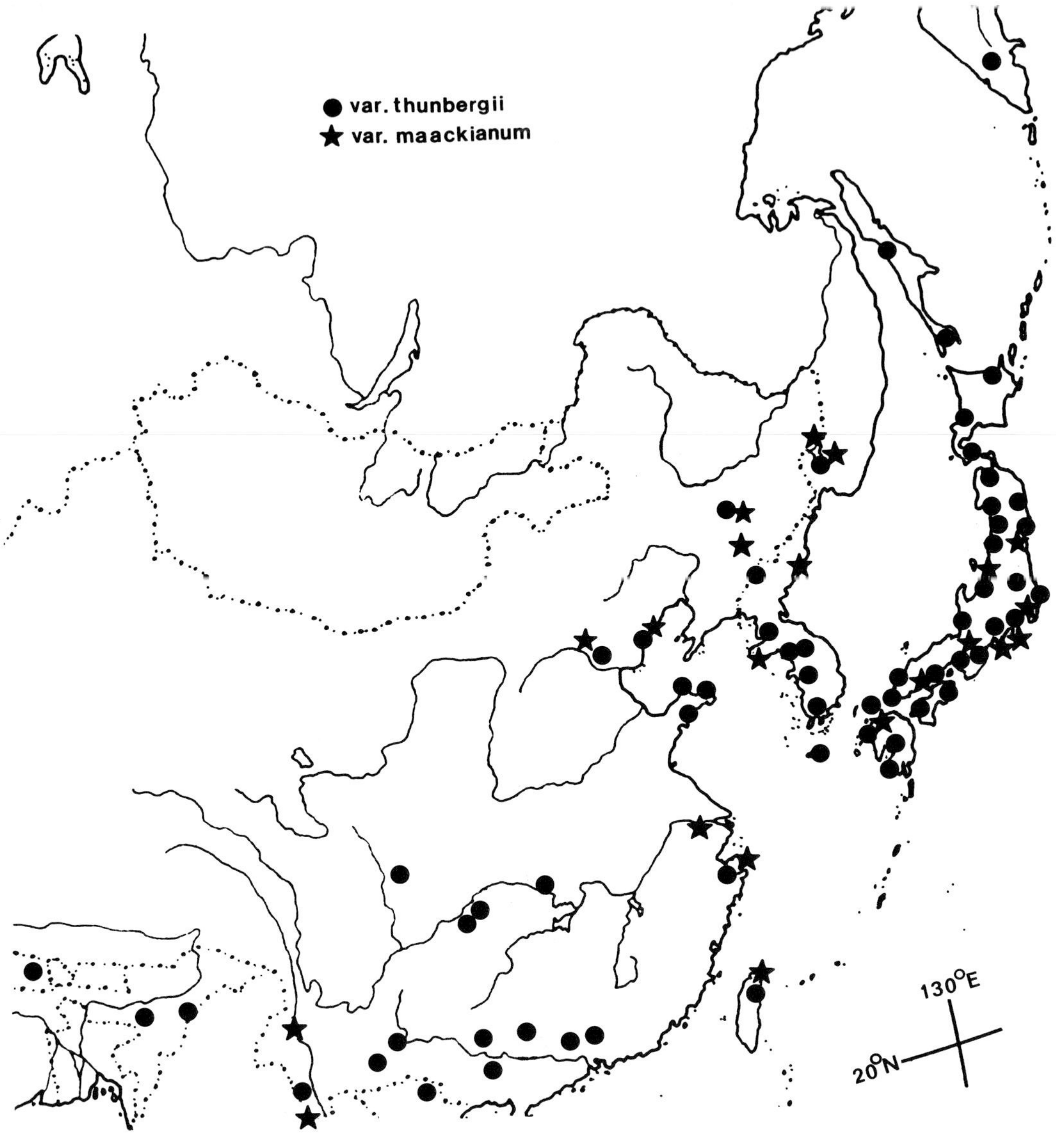

Fig. 16. Distribution of *P. thunbergii.*

(GH). Kwangsi: Tsin Hung Shan, N Hin Yen, under forest, 4000 ft, 19 Aug 1928, *Ching 7003* (NY). Kwangtung: Way to Ju-feng, Lokchong, 10 Oct 1928, *Ying 1218* (E, K, NY). Kweichow: Pin-fa, in 1909, *Cavalerie 2944* (P); Ta Ho Yen, Fan Ching Shan, damp shaded bank, 850 m, 7 Sep 1931, *Steward et al. 384* (NY); Hwei Hsiang Ping, Fan Ching Shan, 2 Oct 1931, *Steward et al. 509* (F, GH, MICH, NY); Weining, in ditch, 7 Oct 1930, *Tsiang 9130* (K, NY, US). Peking: Peking, in 1975, *Chow 75075* (A, MICH, UNC). Shantung: Ming Sha Tung, Lao Shan, along stream, 14 Aug 1930, *Chiao 2945* (F, GH, K, MICH, MT, NY); Miao Tao, 4 Sep 1921, *Licent s.n.* (P); Chefoo, without date, *Col.?* (P). Szechwan: Tchen-keou-tin, without date, *Farges 52* (A, L, P); without specific locality, 1885–1888, *Henry 7258* (NY); Tsien-chuan-fin Hsien, 2000 m, 25 Aug 1930, *Wang 22373* (GH). Yunnan: Without specific locality and date, *Delavay 320* (P); SE Yunnan, Mar-li-po, Hwang-jin-in, on slope, 1100–1400 m, 13 Nov 1947, *Feng 13269* (A); Mi-le?, without date, *Henry 9923* (E, K, NY); Meng-tsze, without date, *Henry 11351* (K, NY, US).

INDIA. Assam: Mawphlang, Khasi Hills, 6000 ft, 11 Sep 1949, *Chand 2221* (L, MICH); Takubama, Naga Hills, ca. 7000 ft, 4 Aug 1950, *Chand 3385* (L, MICH).

JAPAN. Without specific locality, Jun 1877, *Bisset*

303 (LE); without specific locality, 17 Sep 1889, *Faurie 4815* (K); without specific locality, 7 Nov 1893, *Faurie 11406* (P); without specific locality, in 1851, *Göring s.n.* (P); without specific locality and date, *Hogg s.n.* (NY); without specific locality, 1859–1863, *Siebold s.n.* (A); without specific locality and date, *Zollinger 210* (A); without specific locality and date, *Zollinger 3106* (P). AICHI: Minamishidara-gun, Mt. Hooraiji, 200 m, 11 Oct 1962, *Kanai 6867* (TI); Prov. Mikawa, Kitashitara-gun, Inabu-cho, 10 Oct 1967, *Torii s.n.* (A); Prov. Mikawa, Kitashitara-gun, Toyone-mura, Mt. Chausu, 17 Sep 1967, *Torii & Kato s.n.* (UNC). AKITA: Prov. Ugo, Akita-shi, Nibetsu, ca. 100 m, 28 Oct 1980, *Endo 220* (TUS); Akita-shi, in 1927, *Makino 24102* (MAK); Prov. Rikuchu, Waga-gen, Geto, 7 Sep 1927, *Okada s.n.* (TUS). AOMORI: Ravine of the Oreise River, E of Lake Towada, 27 Aug 1928, *Beattie & Kurihara 10334* (US); Aomori, 11 Sep 1885, *Faurie 1132* (K); Kuroishi, 19 Sep 1885, *Faurie 1227* (K); Shimokita-gun, Saimura, Harada, 13 Oct 1956, *Mori s.n.* (MAK 23000); Prov. Mutsu, Sarugamori in Shimokita, 31 Aug 1964, *Yoshioka & Sugawara s.n.* (TUS). CHIBA: Prov. Shimousa, Komuro-machi, Funabashi-shi, 17 Oct 1965, *Ohba 1992* (TI); Mt. Uchiurayama, Awa-kominato-machi, 6–8 Aug 1968, *Ohba et al. 6808061* (TI). EHIME: Prov. Iyo, Matsuyama City, 10 Nov 1954, *Tokui s.n.* (MT). FUKUI: Takefu-shi, Ooshio, 60–160 m, 11 Oct 1976, *Kurosaki 8012* (TI). FUKUOKA: Mt. Seburi, 4 Nov 1934, *Hatusima s.n.* (US); Tsukushi, Dazaifu, foot of Mt. Honam, 23 Sep 1924, *Ichikawa 81* (GH); Kasuya-gun, Hakozaki, 23 Oct 1928, *Ichikawa 72* (BM, P); Kasuya-gun, Hakozaki, 2 m, 1 Nov 1928, *Ichikawa 1* (L, NY, US). FUKUSHIMA: Tamura-gun, E foot of Mt. Ohtakine, on sandy soil by stream, 8 Sep 1968, *Koyama & Kiguchi 2912* (A, DAO, E, L). GIFU: Gifu-shi, 6 Oct 1981, *Endo 403* (TUS); Prov. Mino, 22 Oct 1933, *Shiota 7060* (GH); Prov. Mino, 25 Sep 1934, *Shiota 7894* (GH). GUNMA: Tone-gun, Ozenuma, Aug 1904, *Makino 17420* (MAK); Takasaki-shi, Kwannon-yama, 17 Jul 1966, *Ohba s.n.* (TI); Kurotaki-yama, 23 Sep 1956, *Satomi s.n.* (MAK 1858). HOKKAIDO: Hakodate, in 1857, *Barthe s.n.* (P); Forest reservation nr. Nopporo, Sep 1926, *Dorsett & Morse 1044* (US); Sapporo, 12 Sep 1888, *Faurie 3136* (K); Sapporo, 12 Sep 1888, *Faurie 3144* (K); Prov. Ishikari, Nopporo, 3 Sep 1915, *Kudo s.n.* (TUS); without specific locality and date, *Maries s.n.* (GH, K, NY); Abashiri-gun, Uesato, 8 Sep 1945, *Matsuki 215* (MAK); Hakodate, in 1861, *Maximowicz s.n.* (GH, P, US); without specific locality and date, *Pierot 492* (L); Sapporo, Jul 1880, *Col.?* (GH). HYOGO: Kobe, 31 Aug 1940, *Esoda s.n.* (SP); Kanzaki-gun, Ohkawauchi-cho, Hase, 16 Oct 1966, *Iwatsuki & Koyama 797* (L, MAK, MSC, NY); Ono-shi, E foot of Aonogahara Hill, ca. 40 m, 10 Oct 1970, *Seto 18848* (NLU). IWATE: Mizusawa-cho, 27 Sep 1930, *Iwabuchi 1913* (TUS). KAGOSHIMA: Prov. Satsuma, 15 Oct 1922, *Masamune s.n.* (NY); Prov. Osumi, Ins. Yakushima, 17 Oct 1965, *Sugaya et al. s.n.* (TUS). KANAGAWA: Prov. Sagami, Kamakura, 27 Apr 1852, *Asai s.n.* (TI); Yokohama-shi, in 1911, *Makino 17446* (MAK); Kawasaki City, Suge, 11 Nov 1923, *Makino 17412* (DAO); Kawasaki City, Noborito, in 1931, *Makino 17252* (E); Hakone, 22 Sep 1932, *Makino 17443* (DAO, MAK); Yokohama, in 1862, *Maximowicz s.n.* (GH, NY, P); Yokoska, 1866–1874, *Savatier 1037* (P, US); "circa Yokoska," without date, *Savatier s.n.* (K, P); Sagami, 3 Oct 1934, *Shiota 7948* (GH); Yokohama, in 1862, *Siebold s.n.* (LE); Zushi-shi, Ginmugi, 15 Dec 1968, *Tateishi s.n.* (TUS). KUMAMOTO: Kawachi-toge, in 1911, *Makino 17437* (MAK); Higo, Mt. Nokeyeboshi, 8 Oct 1939, *Mayebara s.n.* (TI); Higo, Yamae, 26 Sep 1960, *Mayebara s.n.* (TI); Higo, Yumotsuru, 28 Oct 1958, *Sohma 5810218* (TUS). KYOTO: Yamabana, 26 Oct 1930, *Kitamura s.n.* (MICH). MIE: Uziyamada-shi, Uziyamada, in 1924, *Makino 17404* (MAK). MIYAGI: Shichiga-shuku-machi, Watarase, on S side of the river, 10 Oct 1978, *Aiba s.n.* (TUS); Rikuzen, Murata-machi, Mt. Tani-yama, 14 Oct 1975, *Fujimaki s.n.* (TUS); Rikuzen, Miyagi-cho, Yatsumori, 8 Oct 1970, *Kimura s.n.* (TUS); Rikuzen, Miyatojima, 13 Oct 1956, *Kyogoku s.n.* (TUS); Sendai-shi, 4 Oct 1972, *Naito 72104* (E, TI, UNC); Ishinomaki-shi, Magiyama, 15 Oct 1972, *Naito s.n.* (TUS); Ishimaki-shi, Nangakusan, 12 Sep 1905, *Odazima s.n.* (MAK 17317); Rikuzen, Masuzawa, Daiwa-cho, 9 Oct 1972, *Ogura 220* (TUS); Rikuzen, Okunikkawa, 2 Oct 1960, *Ohashi s.n.* (TUS); Prov. Iwaki, Watari, Yukiana, 20 Oct 1969, *Saito s.n.* (TUS); Sendai-shi, Aobayama, 15 Oct 1974, *Takahashi et al. s.n.* (TUS). NAGANO: Asama-yama, in 1882, *Hancock s.n.* (K); Ikeda-cho, Aizome-mura, 30 Sep 1922, *Makino 17430* (MAK); Chiisagata-gun, Sanada-cho, Sugadaira, swampy place, 1200 m, 24 Aug 1965, *Murata & Koyama 78* (MAK). NAGASAKI: Prov. Hidzen, Ins. Hirato, 15 Nov 1950, *Kitamura s.n.* (US); Nagasaki, in 1862, *Oldham 694* (GH, L). NARA: Nara-shi, Kasugayama, Oct 1921, *Makino 17405* (MAK); Nara-shi, Kasugayama, 14 Sep 1924, *Michitami s.n.* (GH). NIIGATA: Murakami-shi, Miomotegawa, 13 Sep 1936, *Kamakura s.n.* (MAK 17427); Niigata-shi, 1 Oct 1905, *Okamura s.n.* (MAK 17381). OKAYAMA: Atetsu-gun, 22 Sep 1948, *Uno s.n.* (A); Prov. Bitchu, Kawakami-gun, Chikani, Oct 1914, *Yoshino s.n.* (TI); Prov. Bitchu, Jyobo-gun, Sayodani-mura, Sep 1915, *Yoshino s.n.* (TI). OSAKA: Kawachi, without date, *Yoshida 5561* (GH). SAITAMA: Prov. Musashi, Kobotoke, 17 Oct 1934, *Hisauti s.n.* (TI); Prov. Musashi, Tachibana-gun, Nagatsuta, 4 Nov 1951, *Mizushima 1486* (A); Koshigaya in Musashi, 12 Nov 1950, *Ohwi 170* (DAO, F, K, L, MICH, MT, NY, US); Sekido in Musashi, 24 Oct 1937, *Okuyama s.n.* (DAO); Setagaya, 8 Oct 1893, *Watanabe s.n.* (GH). SHIGA: Shiga-gun, Katata-cho, Hanaore-toge, wet place nr. streams, 27 Sep 1964, *Murata 19029* (L, MSC, NY, US); Ohtsu-shi, Mt. Hiei-zan, 15 Sep 1982, *Ohashi et al. 8860* (TUS). SHIMANE: Okinoshima, 13 Aug 1889, *Hori s.n.* (MAK 17434); Nita-gun, Yokota-machi, 9 Sep 1980, *Ohba & Akiyama 2145* (TI). SHIZUOKA: Fuji-yama, 20 Aug 1890, *Faurie 6648* (P). TOCHIGI: Nikko, Hanaishi, Jyakkonotaki, 660–900 m, 7 Sep 1978, *Ohashi & Ohba 456* (TUS); Nikko, 4 Aug 1912, *Takeda s.n.* (K). TOKUSHIMA: Mt. Ryuozan in Awa, 2 Oct 1960, *Inobe s.n.* (K); Mima-gun, Kenzan, 11 Oct 1908, *Kasai s.n.* (MAK 17435); Tsurugi-san, 1700 m, 10 Aug 1976, *Yamazaki 1092* (TI). TOKYO: Kamiitabashi, Itabashi-ku, 29 Oct 1904, *Makino 17411* (L); Itabashi-ku, Shimura, Nov 1909, *Makino 17414* (DAO, MAK); Itabashi-ku, Shimura, Nov 1909, *Maki-*

no 17451 (L, MAK); Kita-ku, Tabata, 26 Oct 1910, *Makino 17452* (DAO); Ota-ku, Kamata, in 1923, *Makino 17409* (L); Toshima-ku, Ikebukuro, in 1923, *Makino 17410* (L); Shibuya-ku, in 1924, *Makino 17413* (L); Musashino City, Inogashira Park, 8 Oct 1933, *Makino 17448* (DAO, LE); Nerima-ku, Kamishakujii, Oct 1934, *Makino 17450* (L); without specific locality, 29 Sep 1879, *Matsumura s.n.* (US); Osawa, Mitaka-machi, 10 Nov 1946, *Sakamoto 152* (US); nr. Hikawa, 14 Oct 1939, *Suzuki 261* (A). YAMAGATA: Prov. Uzen, Mt. Atsumi-dake, 25 Sep 1951, *Hurusawa 1419* (K); Higashine City, Mt. Shirahige, 13 Oct 1972, *Ogura 183* (TUS); nr. Atsumi Village, along roadside, 25 Sep 1951, *Walker 7344* (US). YAMAGUCHI: Miwa-cho, Tateiwa, 13 Oct 1956, *Wada s.n.* (MAK 100383). YAMANASHI: Prov. Kai, Minami-tsuru-gun, Nakano-mura, Hirano, in shore of Yamanaka Lake, 980 m, 23 Sep 1958, *Furuse s.n.* (A); shore of Yamanaka Lake, Hirano, 22 Sep 1959, *Furuse s.n.* (A); Kiyosato-mura, Yatsuga-dake, in 1936, *Makino 17483* (MAK).

KOREA. "Korea septentrionalis," 11 Aug 1897, *Komarov s.n.* (K, NY); "Korea septentrionalis," 19 Aug 1897, *Komarov 554* (GH). CHEJU: Cheju Isl., 28 Sep 1908, *Taquet 1306* (E, K); Cheju Isl., 29 Oct 1908, *Taquet 1301* (K); Cheju Isl., Seo-gui-po, 22 Jul 1954, *Chung 15303* (DUKE). CHUNGBUK: Mt. Sokni, 21 May 1936, *To s.n.* (SNU). CHUNGNAM: Kongju, 13 Sep 1931, *Chung s.n.* (SNU). CHUNNAM: Mt. Chiri, 10 Aug 1934, *Smith s.n.* (MICH). KANGWON: Mt. Sulak, 14–15 May 1962, *Y. Chung s.n.* (DUKE); Mt. Keonbong, 18 Sep 1972, *Y. Chung s.n.* (SNU); Chuncheon, Deungsun-pokpo, 23 Aug 1969, *Migo s.n.* (TUS); Yong-pyung, Mt. Balwang, nr. ditch, 2 Aug 1981, *Park 1010* (BH); Yong-pyung, Mt. Balwang, wet place along stream, 2 Aug 1981, *Park 1011* (BH); Mt. Kumkang, 26 Aug 1917, *To s.n.* (SNU). KYUNGGI: Kwang-reung, 15 Oct 1947, *Chung 3565* (MICH). PYUNGYANG: Pyungyang, Aug 1930, *Dorsett & Morse 6319* (NY, US); Pyungyang, Jul 1934, *Smith s.n.* (F). SEOUL: Chang-kyung Palace, 14 Sep 1940, *Chung s.n.* (SNU); Biwon, 14 Aug 1940, *Chung s.n.* (SNU); Tae-neung, 29 Aug 1944, *Chung s.n.* (SNU); Mangwoori, 2 Sep 1949, *Chung 9244* (F, MICH); without specific locality, 2 Sep 1914, *Mills 947* (TI); Noryangjin, 8 Sep 1934, *Col.?* (MICH).

NEPAL. East Nepal, Mai Majuwa-Mai, Pokhari-Dhara Pani, 4 Dec 1963, *Hara et al. s.n.* (A).

TAIWAN. Without specific locality, in 1935, *Makino 17400* (MAK). ILAN: Chihtuan, 10 Jun 1977, *Kuo 8738* (TAI).

U.S.S.R. KAMCHATKA: The hot spring at Savav River?, 30 Jul 1921, *Hultén 2510* (GH, NY); without specific locality, 25 Aug 1908, *Komarov s.n.* (P). MARITIME TERRITORY: Dist. Primorskaja oblast, ad. opp. Nikolsk-Ussurisk, 23 Jul 1918, *Kiss s.n.* (MICH). SAKHALIN: Korsakof, 26 Aug 1908, *Faurie 631* (E); northern Sakhalin, 24 Aug 1922, *Kudo s.n.* (TUS); Aleksandrovsk, 12 Sep 1923, *Kudo s.n.* (TUS).

VIETNAM. Tonkin, Chapa, Jul 1924, *Pételot 1622* (GH); Tonkin, Chapa, 1500 m, 25 Aug 1941, *Pételot 6926* (P).

4b. Polygonum thunbergii var. **maackianum** (Regel) Maxim. ex Franchet & Savat., Enum. pl. Jap. **2:** 475. 1878.

Polygonum maackianum Regel, Mém. Acad. Imp. Sci. Saint Pétersbourg, sér. 7. **4(4):** 127, tab. 10, figs. 1 & 2. 1861. *Tracaulon maackianum* (Regel) Greene, Leafl. Bot. Observ. Crit. **1:** 22. 1904. *Persicaria maackiana* (Regel) Nakai ex Mori, Enum. pl. Cor. 132. 1922. *Truellum maackianum* (Regel) Sojak, Preslia **46:** 146. 1974. Type. U.S.S.R. "Am Sungatschi und Kengka-See zwischen Schilf an den Ufern," *Regel s.n.* (holotype, LE!).

Plant moderately to densely stellate-pubescent throughout. *Leaves* smaller, strongly hastate with two widely divergent basal lobes, 4.5–7.5 cm long, 1–2.3 cm wide (middle lobe); middle lobe narrower, lanceolate to narrowly elliptic, acute to acuminate at the apex; basal lobes narrowly triangular to lanceolate, subobtuse to acute; both surfaces usually moderately to densely stellate-pubescent, rarely glabrous; petioles 2.5–5 cm long. *Ocreae* bearing expanded orbicular foliaceous limbs; limbs 0.7–2 cm in diameter, conspicuously dentate; each tooth tipped by a strong bristlelike hair. *Inflorescence branches* sparsely hispid with bristlelike prickles, moderately to densely stellate-pubescent, also with a few glandular hairs in upper portion; bracts and bracteoles usually densely stellate-pubescent. *Achenes* dark brown to black, almost smooth, shiny.

Distribution. From Japan and Korea, through eastern Siberia and Manchuria, to eastern China (Fig. 16).

Additional specimens examined. CHINA. CHEKIANG: Ningpo, Aug 1887, *Faber s.n.* (K). HOPEH: Chihli Prov., Tang Ho, Pei Niu Ting Mt., 5 Aug 1913, *Clemens 8052a* (E); vic. Peitaiho, side of stream, sand dunes, in 1919, *Cowdry 344* (US). KIANGSU: Nanking, 25 Sep 1920, *Merrill 228* (E, MT); Chinkiang, 30 Jun 1934, *Migo s.n.* (NSM). KIRIN: Kirin, 20 Aug 1931, *Chen 376* (GH); "Ad fl. Sungari pr., Sanshan, in herbosis ripariis," 31 Jul 1903, *Desoulavi 2844* (LE, MICH, US); Sungari River valley, 1 Oct 1937, *Skvortzov s.n.* (GH). LIAONING: Liao-tung Peninsula, nr. Sta. Vanfangoo, 21 Jul 1902, *Litvinov 1708* (NY). YUNNAN: Hills around Tengyuek, 6000 ft, Sep 1924, *Forrest 25120* (BH, E, K, NY); Fo-hai, Jun 1936, *Wang 74400* (A).

JAPAN. CHIBA: Ichigawa-shi, Mama, in 1902, *Makino 14992* (LE, MAK); Ichigawa-machi, Pond Junaiike, in 1930, *Makino 14991* (DAO); Imbanuma in Shimousa, 15 Oct 1939, *Okuyama s.n.* (BH); Prov. Shimousa, 25 Sep 1911, *Col.?* (US). FUKUOKA: Yame, Fukushima, 5 Sep 1911, *Seo s.n.* (MAK 14983). GIFU: Mino Prov., 5 Aug 1919, *Shiota 3769* (GH). KANAGAWA: Yokohama, Totsuka, Sep 1905, *Makino 14985* (MAK); Yokohama, in 1862, *Maximowicz s.n.* (K); Yokoska, 1866–1874, *Savatier 2149* (P). MIYAGI: Rikuzen, Shichigo-mura, Onuma, 25 Sep 1931, *Kimura & Naohara s.n.* (TUS). NIIGATA: Nisiurahara-gun, Sirone-mura, 13 Sep 1904, *Kurihara s.n.* (MAK 14979). OKAYAMA: Okayama-shi, Mikado, 24 Sep 1913, *Uno s.n.* (MAK 14981). SAITAMA: Kochitani-shi, Sakurai-mura, Noda, 5 Jul?, *Ito s.n.* (MAK 14975); Prov. Musashi, 26 Sep 1926, *Masamune s.n.* (NY); Koshigaya in Musashi, 12 Oct 1950, *Ohwi 25* (BM, DAO, E, F, L, MICH, MT, NY, TUS, US); Toda, 26 Sep 1937, *Togashi s.n.* (TI). SHIZUOKA: Prov. Suruga, 27 Aug 1927, *Hisauti s.n.* (TUS). TOKYO: Tokyo, Koiwa, 2 Oct 1904, *Makino 14993* (E, US); Itabashi-ku, Shimura, in 1924, *Makino 14994* (L, MAK); Katsuura-ku, Mizumoto, Sep 1938, *Makino 14989* (MAK); Ohizumi-machi, 17 Sep 1940, *Makino 14987* (DAO).

KOREA. HAMBUK: Eo-dae-jin, 19 May 1935, *Nakai s.n.* (SNU). HWANGHAE: Changsangod, 29 Jul 1929, *Nakai 12708* (TI); Changsangod, 29 Jul 1930, *Nakai s.n.* (SNU).

TAIWAN. Without specific locality, in 1864, *Oldham 431* (GH, K, LE, P). TAIPEI: Taipei, 24 Oct 1926, *Sasaki 75* (MICH); Taipei, 9 Oct 1933, *Tanaka & Shimada 17863* (E, F, GH, MICH, NY, P, TI).

U.S.S.R. MARITIME TERRITORY: Dist. Primorskaja oblast, ad opp. Nikolsk-Ussurisk, 10 Sep 1920, *Kiss s.n.* (MICH); Ussuri, without date, *Maack s.n.* (P).

5. **Polygonum biconvexum** Hayata, J. Coll. Imp. Univ. Tokyo **25:** 184. 1908. *Polygonum hastatotrilobum* var. *lenticulare* Danser, Bull. Jard. Bot. Buitenzorg, sér. 3. **8:** 227. 1926. *Persicaria biconvexa* (Hayata) Nemoto, Fl. Jap. Suppl. 169. 1936. *Truellum biconvexum* (Hayata) Sojak, Preslia **46:** 145. 1974. Type. Taiwan. "In monte Morrison," in 1905, *Nakahara s.n.* (not seen; see Park, Brittonia **38:** 398. 1986).

Ascending *herb* arising from long decumbent base, 20–80 cm tall; stem branched, angled, sparsely to moderately armed with recurved prickles, often with minute 6- to 8-rayed sessile stellate hairs; prickles stout, 1–1.5 mm long; rays of the stellate hairs very short, somewhat ascending; lower nodes usually producing roots. *Leaves* long-petioled; blades broadly triangular in outline, 1.8–4.7 cm long, 1.8–3.6 cm wide, acuminate at the apex, hastate with two somewhat divergent lobes at the base, strongly ciliate at the margin; basal lobes small, triangular to ovate, acute to obtuse; middle lobe broadly elliptic to ovate, usually contracted near the base; both surfaces moderately to densely pubescent with appressed bristlelike simple multiseriate hairs to 2 mm long intermixed with varying densities of 4- to 7-rayed sessile stellate hairs; rays slender, much longer than those of stem; midveins beneath retrorsely prickly; petioles 1.5–2.6 cm long, usually antrorsely prickly with bristlelike prickles to 1.5 mm long, often with stellate hairs identical to those of stem in upper portion. *Ocreae* short, 0.5–1 cm long, cylindrical, truncate at the apex, usually scarious but often with a small foliaceous limb, long-ciliate. *Inflorescences* terminal or in the axils of upper leaves, elongate, 3–4 cm long, often branched, sparsely hispid with bristlelike prickles to 1.5 mm long, sparsely to moderately glandular-pubescent; glandular hairs stalked, ca. 0.5 mm long; inflorescence bracts resembling ocreae but much reduced, 2–3 mm long, often associated with reduced leaf; each inflorescence branch terminated by a dense headlike cluster of many flower fascicles. *Flower fascicles* usually 2-flowered; bracts lanceolate, acuminate at the apex, glabrous, usually long-ciliate, 3.5–4 mm long; bracteoles elliptic, 3–3.5 mm long, glabrous, ciliate at the tip; pedicels short, 2–3 mm long, glabrous, included in bracteoles; perianth 4-parted, whitish-pink to reddish, 3–4 mm long; tepals elliptic, obtuse, not becoming fleshy; stamens usually five, rarely six, inserted at the base of perianth in two whorls; inner stamens two, 1.5–2 mm long; outer stamens 3–4, 1–1.5 mm long; style one, deeply 2-cleft, often slightly exserted; stigmas two. *Achenes* biconvex, obovate in outline, pale brown to grayish-brown, acute at the apex, somewhat rugose, shiny, ca. 3 mm long.

Distribution. Known only from Taiwan and Sumatra (Fig. 17).

Polygonum biconvexum is very closely related to *P. thunbergii* but is distinct from the latter in having a four-parted perianth, a deeply two-cleft style, and biconvex achenes. In addition, leaf surfaces of *P. biconvexum* are usually densely pubescent with appressed bristlelike simple multiseriate hairs.

Additional specimens examined. INDONESIA. SUMATRA: Kerintji, 700 m, 3 Oct 1920, *Bünnemeiyer 8932* (L); Kerintji, Soengai, Koembang, 1400 m, Sep 1915, *Jacobson 2522* (L); Korinchi, Sungei, Kumbang, Apr 1914, *Robinson & Kloss s.n.* (K).

TAIWAN. Without specific locality and date, *Faurie s.n.* (P); without specific locality, 7 Aug 1919, *Matuda s.n.* (TI). CHIAYI: Mt. Alishan, 2100–2400 m, 10 Aug 1969, *Ando et al. 476* (TUS); Mt. Alishan, roadside, 21 May 1966, *Chuang 4207* (L, TI); Mt. Alishan, 2150 m, May–Aug 1934, *Gressitt 221* (GH, L); Mt. Alishan, in 1912, *Hayata s.n.* (TI); Mt. Alishan, mountain slope in an artificial *Cryptomeria* forest, 20 Oct 1970, *Hsu 8856* (TAI); Mt. Alishan, 15 Sep 1960, *Huang & Kao 1607* (L, MO); Mt. Alishan, Dec 1933, *Kanehira 3007* (NY, US); Mt. Alishan, shaded mountain slope, 2700 m, 17 Aug 1957, *Ream 604* (GH). ILAN: Taipingshan, 7 Aug 1928, *Suzuki s.n.* (US); Taipingshan, 27 Jul 1929, *Suzuki s.n.* (MICH). KAOHSIUNG: Liukuei, Mt. Peinan-chushan along Lao-nung-chi, 1800 m, 30 Aug 1969, *Yamazaki et al. s.n.* (TI). NANTOU: Chitou, 7

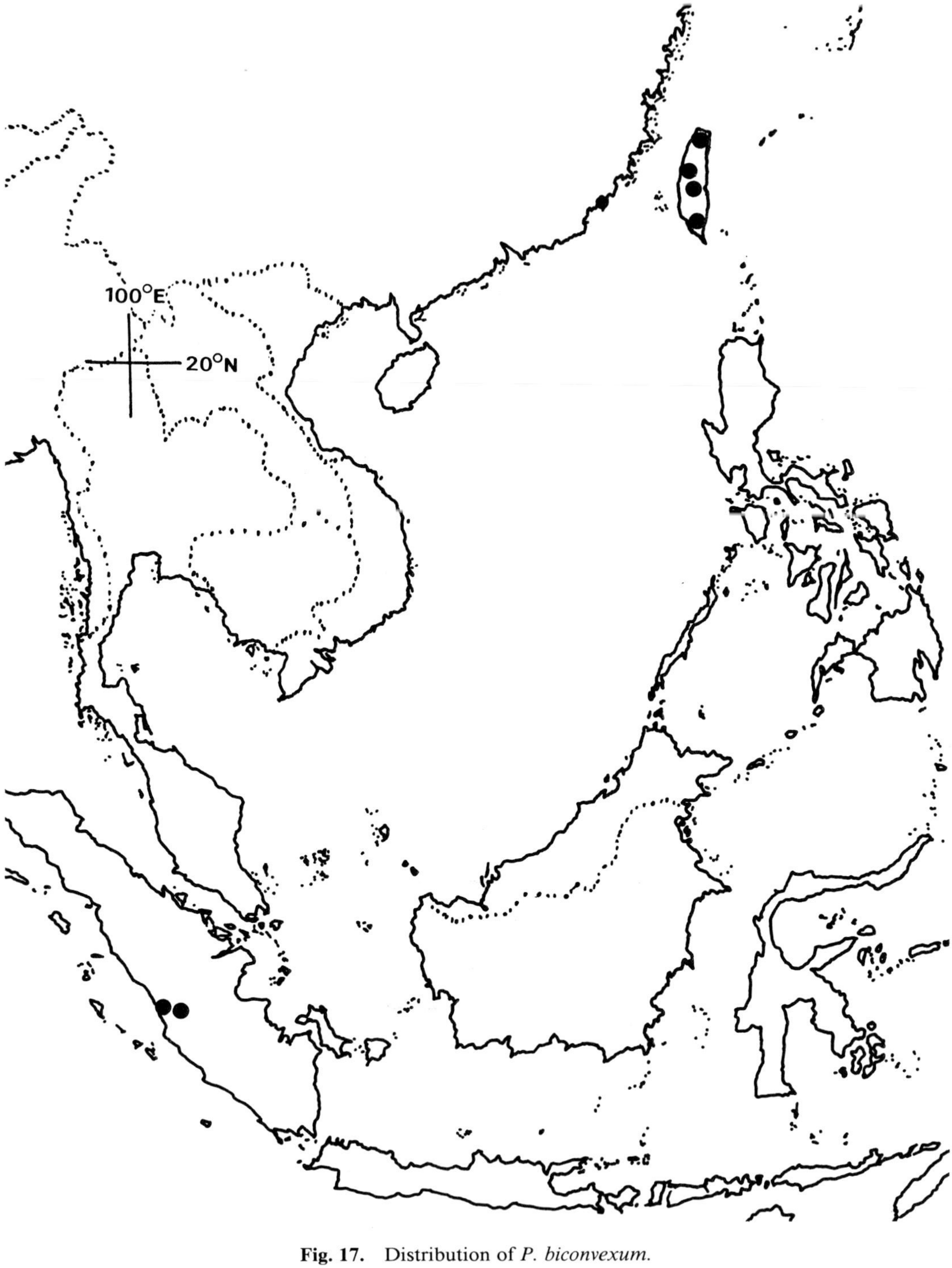

Fig. 17. Distribution of *P. biconvexum.*

Sep 1981, *Chen 2763* (TUS); Chitou, roadside, 6 Oct 1965, *Chuang 4004* (L); Chitou, roadside, 8 Oct 1965, *Chuang 4071* (L); Fonghuangshan, 25 Dec 1978, *Kuo 10342* (TAI); Chitou, 1100 m, 8 Oct 1966, *Van Steenis 20767* (L). TAICHUNG: Shun-kuan, along highway, 2 Sep 1973, *Huang & Hsieh 7321B* (TAI); Mt. Ammashan, 10 Oct 1957, *Liu et al. 377* (A). TAIPEI: Taipei, 25 Sep 1930, *Suzuki 6169* (BM). TAITUNG: Chih-pen-wn-chien, 600–1600 m, 27 Jul 1967, *Hsu & Kao 3293* (TAI).

6. **Polygonum arifolium** L., Sp. pl. 364. 1753. *Tracaulon arifolium* (L.) Raf., Fl. tellur. **3:** 13. 1837. *Truellum arifolium* (L.) Sojak, Preslia **46:** 145. 1974. *Persicaria arifolia* (L.) Haraldson, Symb. Bot. Upsal. **22:** 72. 1978. Type. LINN 510.34 [lectotype (microfiche at NY!), designated by Reveal, J. Linn. Soc., Bot. **92:** 168, 1986; without locality, date, and collector's name].

Polygonum sagittatum var. *pubescens* Keller, Bull. Soc. Roy. Bot. Belgique, Part 2. **30:** 45. 1891. *Polygonum arifolium* var. *pubescens* (Keller) Fern., Rhodora **48:** 53. 1946. Type. U.S.A. New York: Wet grounds nr. Troy, in 1870, *Puissant s.n.* (not seen; see Fernald, Rhodora **48:** 53. 1946).

Polygonum arifolium var. *lentiforme* Fern. & Griscom, Rhodora **37:** 167. 1935. Type. U.S.A. Massachusetts: Southwick, swamp along Great Brook, 9 Sep 1914, *Seymour 251* (holotype, GH!; see Park, Brittonia **38:** 399. 1986).

Scandent annual *herb* arising from a fibrous root system, 0.2–1.5 m tall; stem branched, light green, often reddish-purple at maturity, glabrous, angled, armed with recurved prickles along the angles, often producing roots at lower nodes; prickles stout, ca. 1 mm long. Middle and lower *leaves* long-petioled, triangular to broadly elliptic, 6.5–13 cm long, 6–10.8 cm wide, acuminate at the apex, broadly hastate with two widely divergent acuminate triangular lobes at the base, prickly at the margin; upper leaves small, short-petioled, usually lanceolate, 3.5–6 cm long, 1–2 cm wide, cordate or somewhat sagittate at the base; upper surfaces dark green to green, moderately to densely pubescent with appressed simple multiseriate hairs to 1 mm long and with stellate hairs, rarely glabrous; stellate hairs scattered, sessile, 4- to 5-rayed; lower surfaces light green, covered with 4- to 6-rayed sessile stellate hairs, rarely glabrous; midveins and major veins beneath bearing prickles; prickles 1–1.5 mm long; petioles light green, retrorsely prickly. *Ocreae* scarious, 0.8–1.5 cm long, oblique, usually glabrous but often with bristlelike hairs, bristly-ciliate at upper margin, bearing short prickles at the base. *Inflorescences* terminal or in the axils of upper leaves, elongate, often branched, 2–10 cm long, often associated with reduced leaves, usually densely pubescent with glandular hairs and 4- to 5-rayed sessile stellate hairs; glandular hairs stalked, 0.5–0.7 mm long; each inflorescence branch terminated by a loose interrupted panicle or a headlike cluster of 3–8 flower fascicles. *Flower fascicles* 2- to 4-flowered, usually widely spaced; bracts elliptic, hispid, 5–8 mm long, with elongate acuminate apex; bracteoles broadly elliptic, 2–3 mm long, sparsely stellate-pubescent toward the apex; pedicels short, 2–3 mm long, glabrous, almost included in bracteoles; perianth 4-parted, 5–6 mm long, pink or red, often whitish-green at the base; tepals broadly elliptic, not becoming fleshy; stamens 6(–8), inserted at the base of perianth in two whorls; inner stamens two, ca. 1.3 mm long; outer stamens four, ca. 1 mm long; style one, 2-cleft up to middle, ca. 1 mm long, included; stigmas two. *Achenes* biconvex, acute at the apex, dark brown to black, smooth, shiny, 4–6 mm long.

Distribution. In moist situations, usually in shade; predominantly in margins of swamps, ponds, and tidal marshes along rivers, occasionally in shaded thickets along streams and wet ravines in forests; in eastern North America, from Prince Edward Island to Georgia, and from the east coast westward to Minnesota and Kentucky. Also, known from one isolated locality in Washington State (Klickitat Co., Husum, Mt. Paddo) (Fig. 18).

Polygonum arifolium is highly variable in pubescence. In its southern range, which comprises Virginia, North Carolina, South Carolina, and Georgia, the species consistently displays moderately to densely pubescent ocreae. In addition, it tends to be more pubescent on both surfaces of the leaves. However, most populations from the northern part of its range, from Canada to New York, have completely glabrous ocreae. Also, foliar pubescence is far more variable, from densely pubescent on both surfaces to completely glabrous. In New Jersey and Maryland, where interpopulational variation is high, both these forms are found. However, populations of these forms did not show any significant qualitative differences in flavonoid chemistry (Park, 1987).

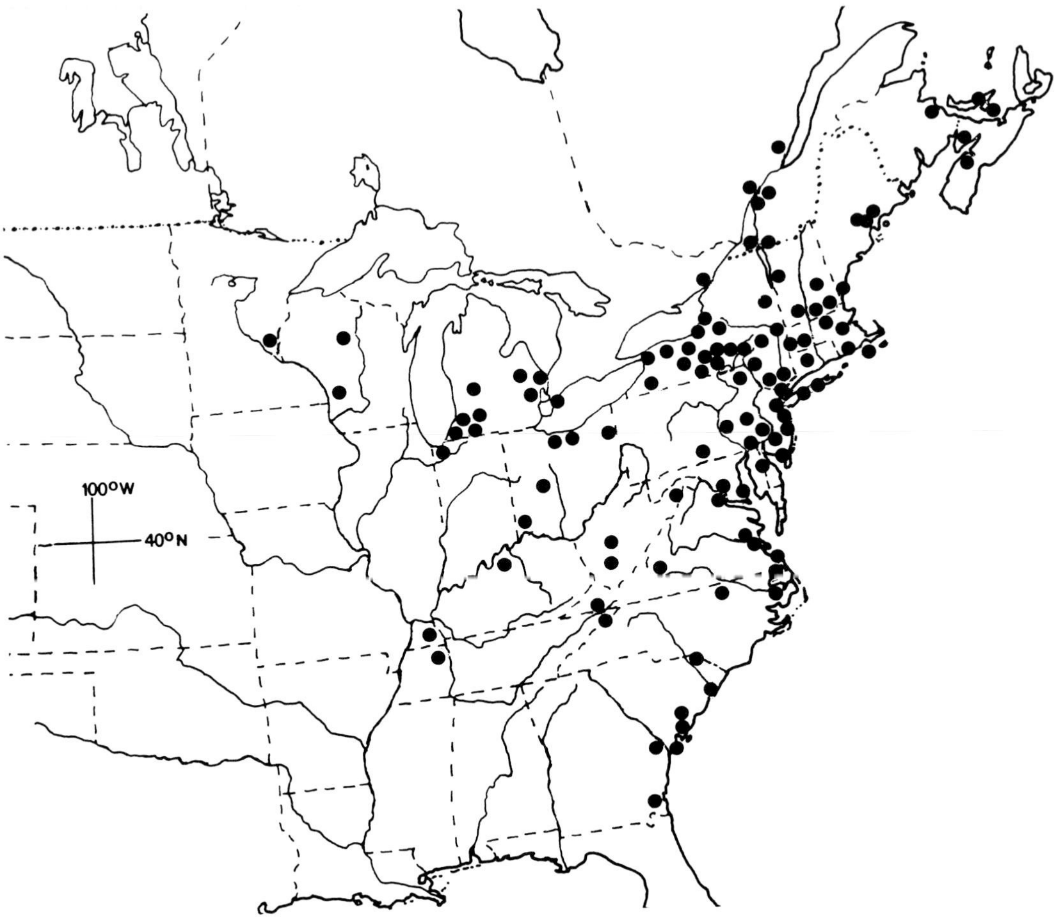

Fig. 18. Distribution of *P. arifolium.*

Additional specimens examined. CANADA. NEW BRUNSWICK. Kent Co.: St. Louis Parish, Kouchibouguac Nat. Park, 0.45 mi S of Frigot Rd. in Oliver Rd., 15 Sep 1977, *Munro 1830* (DAO); St. Charles Parish, Aldouane Peninsula, 3.4 mi E of St. Louis, 18 Jul 1978, *Munro 2023* (DAO). NOVA SCOTIA. Colchester Co.: Economy, wet alder swamp, 31 Aug 1927, *Prince 743* (DAO). Kings Co.: W of Kentville, edge of swamp, 25 Jul 1942, *Dore & Roland s.n.* (DAO). ONTARIO. Kent Co.: 14 km SE of Wheatley, Wheatley Provincial Park, Romney Township, in 1893, *Brunton et al. s.n.* (MICH). Leeds Co.: Front of Leeds and Landsdowne Township, Landon Bay, 17 Aug 1981, *Sutherland & Lawley 1919* (DAO). PRINCE EDWARD ISLAND. Queens Co.: Wood Islands, alder thicket along stream, 26 Aug 1952, *Erskine 1625* (MT, NY); Beach Grove, 2 mi NW of Charlottetown, 2 Aug 1953, *Erskine 2260* (NY); Lake Verde, swampy woods and thickets, 9 Aug 1912, *Fernald et al. s.n.* (MT, US). QUEBEC. Iberville Co.: 2 mi E of Iberville, shaded woods, black mucky soil, 12 Aug 1952, *Bassett & Hamel 2456* (DAO); près de St.-Grégoire, 8 Aug 1952, *Louis-Marie 52600* (DAO); Iberville on the Richelieu River, 26 Aug 1934, *Maire-Anselme s.n.* (CU); Mont Johnson, bois humides, 28 Aug 1936, *Marie-Victorin et al. 46655* (CU, DAO, MT); Saint-Grégoire, 14 Sep 1941, *Marie-Victorin et al. 56081* (MT, NY, US); Mont St. Grégoire, 20 Sep 1944, *Raymond & Kucyniak 2326* (MT). Missisquoi Co.: Philipsburg, 6 Aug 1928, *Adrien 2083* (MT). Montmorency Co.: Ste.-Pétronille, 17 Sep 1946, *Cayouette & Cinq-Mars 46-15* (DAO, MT); Ste.-Pétronille, 6 Aug 1963, *Cinq-Mars et al. 63-1020* (DAO, MT). Napierville-La Prairie Co.: La Prairie, 25 Sep 1934, *Marie-Victorin & Rolland-Germain 43812* (CU, MICH, MT). Nicolet Co.: SS. Assomption, 2 Sep 1949, *Allyre 2761* (DAO). Pontiac Co.: Baie-De-Pontiac, 8 Jul 1931, *Marie-Victorin et al. 43453* (MT). St. Jean Co.: St. Jean, 2 Aug 1953, *Cinq-Mars & Raymond s.n.* (DAO, MT); Talon, bords de la rivière, Jul 1949, *Raymond 690* (MT); Talon, 2 Aug 1953, *Raymond & Cinq-Mars 3014* (DAO, MT). St. Maurice Co.: Pointe-Du-Lac, 23 Jul 1935, *Cléonique 8607* (MT). Richelieu Co.: Sainte-Anne-De-Sorel, 22 Aug 1943, *Marie-Victorin et al. 2295* (DAO, MICH, MT). Rouville Co.: Rougemont, 17 Jul 1954,

Cinq-Mars s.n. (DAO, MT); Rougemont, 20 Jul 1965, *Cinq-Mars et al. 65-207* (DAO); Mont St. Hilaire, 19 Jul 1959, *Maryniak 3087* (DAO).

U.S.A. CONNECTICUT. Hartford Co.: Southington, 27 Aug 1898, *Bissell 182* (NY). New Haven Co.: Wallingford, wet meadow, 14 Sep 1922, *Castle & Dann 34* (CU). DELAWARE. New Castle Co.: Without specific locality, swamps, Aug 1884, *Canby s.n.* (NY); on dredged mud of tidal marshes, along Appoquinimink Creek, S of Odessa, 11 Aug 1934, *Fernald & Long 3920* (GH); marsh nr. Pigeon Point, on river shore SE of Wilmington, 8 Sep 1928, *Tatnall s.n.* (GH). DISTRICT OF COLUMBIA. Rock Creek, 13 Sep 1896, *Pieters s.n.* (MICH); without specific locality, swamp ground, 12 Sep 1896, *Steele s.n.* (GH). GEORGIA. Chatham Co.: Savannah River, swamp nr. Port Wentworth, 25 Oct 1953, *Duncan 17446* (MICH). Camden Co.: In swamp of Satilla River nr. Woodbine, 23 Aug 1902, *Harper s.n.* (GH, NY, US). INDIANA. Porter Co.: Dunes at Keiser, 23 Sep 1923, *Lyon s.n.* (MICH). KENTUCKY. Calloway Co.: Blakusburg, swamp on the W fork of Clarks River on A.C. Hopkins' farm, 15 Sep 1973, *Athey 2574* (UNC, VDB). Henry Co.: 3.2 mi W of rd. junction KY 389 and KY 22, 1.3 mi from hwy., growing in a drained lake, 19 Jul 1962, *Gentry 517* (NY, UNC). MAINE. Penobscot Co.: Orono, shaded margin of small pond, 17 Aug 1908, *Fernald 205* (BH, CU, L, MICH); in coniferous swamp in damp rich soil in eastern Corinna, 23 Aug 1937, *Quimby & Laubengayer s.n.* (CU). Waldo Co.: N shore of Lake Megunticook, wet ravine in forest, 19 Aug 1939, *Friesner 14079* (NY); low area in woods bordering Megunticook Lake nr. mouth of Mullin's Pond, 12 Aug 1950, *Friesner 23694* (GH, MICH, TEX). York Co.: South Berwick, 12 Aug 1933, *Perkins s.n.* (CU). MARYLAND. Calvert Co.: Battle Creek Cypress Swamp Sanctuary, N side of MD 506 just W of Gray Mutual Rd., cypress swamp and woodland along creek banks, 4 Oct 1980, *Hill 9637* (NY); Chesapeake Beach, 27 Aug 1904, *House 381* (NY). Cecil Co.: Open swampy woods along Octoraro Creek, 11 Aug 1923, *Long 28523* (GH). Prince Georges Co.: Ammendale, in 1918, *Hyacinth 1759* (BH). MASSACHUSETTS. Berkshire Co.: East Otis Reservoir, by small stream, 8 Oct 1929, *Manning s.n.* (CU). Bristol Co.: Easton, 9 Sep 1898, *Ames s.n.* (TEX). Dukes Co.: Martha's Vineyard, West Tisbury, along Look's Brook, 22 Aug 1916, *Brown & Seymour 1464* (NY); Martha's Vineyard, Chilmark, 18 Aug 1964, *Mackeever MV656* (NY). Middlesex Co.: West Cambridge, 19 Aug 1905, *Bartlett 260* (MICH); Cambridge, 21 Aug 1862, *Beal s.n.* (BH); Cambridge, 22 Aug 1894, *Blankinship s.n.* (TEX); Natick, 25 Aug 1905, *Chamberlain & Knowlton s.n.* (US); Framingham, 3 Aug 1907, *Eames s.n.* (CU); Arlington, 10 Jul 1884, *Col.?* (BH). Norfolk Co.: Stoughton, 28 Jul 1912, *Blake 4273* (TEX); Stoughton, boggy ground, edge of woods, 7 Sep 1917, *Blake 6740* (TEX); Holbrook, shaded brook, 2 Sep 1918, *Churchill 362* (MICH); Dover, swamp S of Day-Brook Rd., 20 Sep 1912, *Wiegand & Rodman s.n.* (CU). Worcester Co.: Lancaster, wet swampy woods, 15 Aug 1943, *Brown 1785* (TEX); swamp nr. Mendon (Hopedale), 21 Aug 1930, *Manning s.n.* (CU). MICHIGAN. Allegan Co.: In vic. Swan Creek Wildlife Exp. Station, wooded swampland, 23 Aug 1937, *Allen 2* (MICH). Berrien Co.: Rich shaded ground along stream nr. Birchwood Beach, 22 Aug 1917, *Dodge 380* (MICH); Pipeston, 25 Aug 1906, *Gates 1088* (MICH); low ground nr. Pere Marquette Railroad and Rt. 31 not far from the Paw Paw River, NE of Benton Harbor, 20 Sep 1964, *Swink 3265* (MICH). Cass Co.: Without specific locality, margin of a lake, 22 Aug 1838, *Wright s.n.* (MICH). Defiance Co.: Occasional at S end of Little Lake along ditch in woods, Milford Township, ca. 5.5 mi NW of Farmer, 17 Sep 1969, *Stuckey 8566* (MICH). Kalamazoo Co.: N Swartz Swamp, 31 Aug 1933, *Hanes 2333* (MICH). Kent Co.: Grand Rapids, 15 Aug 1891, *Bailey s.n.* (MICH). Lapeer Co.: Edge of Twin Lake, Inslay Township, 30 Aug 1910, *Dodge s.n.* (MICH). St. Clair Co.: Nr. Capac, 24 Aug 1892, *Dodge s.n.* (MICH); Port Huron, 1 Aug 1894, *Dodge s.n.* (NY). Van Buren Co.: South Haven, 25 Jul 1910, *Kauffman s.n.* (MICH); N of South Haven, damp places in ravine, 4 Sep 1911, *Lansing 3337* (US). MINNESOTA. Chisago Co.: Center City, Jul 1892, *Taylor s.n.* (NY); Lindstown, Aug 1892, *Taylor s.n.* (NY, US). NEW HAMPSHIRE. Belknap Co.: Center Harbor, Aug 1891, *Schrenk s.n.* (CU). Cheshire Co.: Keene, Aug 1883, *Gilbert s.n.* (NY). Merrimack Co.: Without specific locality, 13 Sep 1919, *Batchelder s.n.* (NY, US). NEW JERSEY. Bergen Co.: Carlstadt, 30 Aug 1883, *Niederer s.n.* (NY); without specific locality, low grounds, in 1877, *Oehler s.n.* (NY). Burlington Co.: Atsion, edge of swamp, 5 Sep 1917, *Gershoy 289* (CU); Kinkora, 18 Aug 1910, *Taylor 2546* (NY). Camden Co.: West Collingswood, 22 Sep 1922, *Bassett s.n.* (CU, NY). Cumberland Co.: Vic. Seeley's Mill (Cedar Grove) on Cohansey River 4 mi NW of Bridgeton, 1 Sep 1924, *Beals & Bassett s.n.* (NY); Port Elizabeth, on river bank, 20 Sep 1974, *Moldenke 29031* (TEX). Essex Co.: Milburn, 18 Aug 1894, *Livingstone 41446* (MICH); Newark, marshy ground, 28 Aug 1896, *Lyon s.n.* (US). Gloucester Co.: Fries Mills, in 1854, *Ames s.n.* (MICH); wet places ½ mi S of Bridge Port, 23 Aug 1935, *White s.n.* (NY). Hudson Co.: Nr. Hoboken, 18 Aug 1920, *Bailey s.n.* (BH). Hunterdon Co.: Fairmount, 20 Sep 1908, *Dautun? s.n.* (L). Middlesex Co.: Cheesequake State Park, 22 Sep 1940, *Alexander s.n.* (NY); New Durham, 24 Aug 1894, *Kearney s.n.* (NY); Colonia, 20 Aug 1914, *Lighthipe s.n.* (NY); Spotswood, 17 Aug 1910, *Taylor 2487* (NY); Monmouth Junction, 20 Aug 1910, *Taylor 2572* (NY). Monmouth Co.: Belmar, swampy border of Wreck Pond, 8 Sep 1917, *Gershoy 290* (CU); Belmar, wet, rich woodland, 5 Sep 1942, *Thorne 1012* (CU). Ocean Co.: Bay Head, 14 Sep 1914, *Lighthipe s.n.* (TEX). Somerset Co.: Watchung, at edge of pond in Seeley's Notch, without date, *Moldenke 11257* (NY). Union Co.: Watchung Reservoir, along Blue Brook, 17 Aug 1936, *Kezer s.n.* (NY). NEW YORK. Bronx Co.: Low woods N of Bronx Park, 20 Aug 1920, *Burnham s.n.* (GH); Mosholu Parkway, 21 Sep 1902, *Edmondson 2748* (NY); McLean Woods, 8 Sep 1919, *Holtzoff s.n.* (NY). Cattaraugus Co.: Swamp nr. Little Valley, 26 Jul 1927, *Perkins s.n.* (CU). Cayuga Co.: Nr. Duck Lake, Conquest, 12 Aug 1916, *Metcalf 6365* (CU). Chemung Co.: W of Pine City and about 1 mi from the Steuben Co. line, edge of bog, 21 Jul 1945, *Bailey 17275* (BH); Rodbourn Marsh, mucky thickets,

27 Aug 1935, *Smith 351* (CU). Delaware Co.: Margaretville, Karl Brook's Delectable Swamp, 19 Sep 1982, *Brook & Dirig 1187* (BH); Town of Hancock, French Woods, Bouchoux's Swamp, 300–950 ft W of Dirig Rd. and extending from State Rt. 97 on the S, 1400 ft to the N, 19 Aug 1979, *Dirig 385* (BH); Town of Hancock, French Woods, Bouchoux's Swamp, S part of swamp, growing in muck, 19 Jul 1981, *Dirig 766* (BH); Town of Hancock, Pea Brook, *Sphagnum*/heath bog and surrounding swampy woods, 500 ft E of Hungry Hill Rd., 500 ft S of jeep trail and 1150 ft NNW of Delaware Lake, 11 Sep 1977, *Dirig & Cryan s.n.* (BH); Town of Hancock, French Woods, Bouchoux's Swamp, ca. 1770 ft, Sep 1982, *Park & Dirig 809* (BH); Town of Hancock, Pea Brook, old beaver pond, marsh and surrounding swampy woods, 250–750 ft W of Nevins Rd., 300–850 ft E of State Rt. 97, and extending from Cross Rd. on the S, 1400 ft to the N, Sep 1982, *Park & Dirig 810* (BH). Dutchess Co.: Vic. Clove, swamp, 26 Aug 1915, *Standley & Boliman 12106* (US). Erie Co.: Buffalo, without date, *Clinton s.n.* (MICH). Genesee Co.: Town of Pembroke, 4 Sep 1930, *Zenkert s.n.* (CU). Greene Co.: Schoharie Creek, in 1830, *Col.?* (NY). Jefferson Co.: Marshes bordering ponds along Lake Ontario, nr. Woodville, 14 Aug 1932, *House 19867* (TEX). Monroe Co.: Swamp, Durand-Eastman Park, Rochester, 19 Sep 1909, *Edson & Bishop s.n.* (CU). Nassau Co.: Edge of swale, SW of Hempstead Village, Hempstead, 16 Sep 1917, *Gershoy 905* (CU). Oneida Co.: Wooded swamp, Poppelton Rd., Verona Township, E end of Oneida Lake, 9 Sep 1934, *House 22248* (NY); in swamp ¾ mi E of NE corner of Oneida Lake, 17 Sep 1922, *Muenscher 114710* (CU). Orange Co.: Hamptonburgh, 10 Aug 1912, *Denslow s.n.* (NY); in swamp below *Sphagnum* pond, Black Rock Forest, 14 Jul 1936, *Raup 7584* (MICH, NY); West Point, without date, *Col.?* (NY). Oswego Co.: Volney, Birdsall's Bog and vic., 19 Aug 1902, *Rowlee & Wiegand s.n.* (CU); Oswego, 22 Aug 1886, *Stewart s.n.* (CU); swamp bordering high land, 1.5 mi N of Selkirk, 21 Aug 1920, *Wiegand 13515* (CU). Queens Co.: Flushing, 29 Aug 1888, *Northrop s.n.* (NY). Rensselaer Co.: Brunswick, in 1843, *Snyder s.n.* (CU). Richmond Co.: Without specific locality, 3 Aug 1879, *Hollich s.n.* (NY); Bull's Head, 22 Sep 1917, *Wilson s.n.* (NY). Rockland Co.: Rockland Lake, 18 Aug 1923, *House 9823* (CU); S side of Rt. 59 W of Middletown Rd., Clarkestown Township, 9 Sep 1962, *Lehr 414* (NY); Bear Mountain Interstate Park, in swampy ground, 16 Jul 1939, *Moldenke 11197* (NY). Schuyler Co.: In damp woods nr. outlet of Cayuga Lake, 12 Sep 1940, *Laubengayer s.n.* (CU); Patrick Swamp, 7 Aug 1892, *Wiegand s.n.* (CU). Seneca Co.: Seneca Falls, Montezuma Marshes, swampy woods, 7 Aug 1918, *Eames 9844* (CU); Tyre, swamp along State Rd. through Montezuma Marshes, 9 Sep 1919, *Eames et al. s.n.* (CU). Suffolk Co.: Cold Spring Harbor, 14 Aug 1935, *Cain 534* (NY); Hauppauge, 17 Aug 1936, *Cain s.n.* (NY); outlet of Belmont Lake, N of Babylon, without date, *Muenscher & Curtis 6040* (CU, NY); east margin of Jarvis Pond, 5 Aug 1904, *Shull 221* (NY). Tioga Co.: Headwater's Swamp, 8 Sep 1982, *Cope & Dirig 121* (BH). Tompkins Co.: Ithaca Flats, 28 Sep 1872, *Dudley s.n.* (CU); Dryden, Freeville and vic., valley of Fall Creek, low woods, Fir-tree Swamp, 8 Jul 1918, *Eames & Gershoy 9843* (CU); Caroline, Slaterville Swamp, 13 Jul 1919, *Eames et al. s.n.* (CU); Danby, Michigan Hollow, swampy woods, 23 Jul 1919, *Eames & Wiegand 11990* (CU); Ithaca, valley of Cayuga inlet, 23 Aug 1916, *Metcalf & Wiegand 6366* (CU); Round Marshes (Gracie's Swamp) and vic., Alder Swamp, NE of Mud Pond, 28 Sep 1924, *Muenscher & Wiegand 15443* (CU); Ithaca, open alluvial and marshy flats between Fall Creek, East Hill and Cayuga Lake, 25 Aug 1913, *Palmer 419* (CU); margin of McLean Bog, 6 Oct 1981, *Park 705* (BH); Ithaca, 23 Sep 1894, *Wiegand s.n.* (BH). Warren Co.: Marsh at Brayton nr. Lake George, 22 Jul 1942, *House 28824* (CU). Washington Co.: East Greenwich, in 1867, *Fitch s.n.* (MICH); Pilot Knob, east side of Lake George, shaded marshy soil, 9 Oct 1943, *House 28858* (CU). Westchester Co.: Above Bronxville, 30 Sep 1893, *Bicknell s.n.* (NY); Pelham Manor, 5 Aug 1878, *E.H.D. s.n.* (NY); White Plains, 7 Aug 1915, *Wilson s.n.* (NY). Yates Co.: In Five Mile Swamp, 14 Jul 1952, *Muenscher & Schumacher 23527* (CU); Potter Swamp, thick wet woods along Nettle Valley Creek, 1 mi E of Potter, 880 ft, 8 Oct 1950, *Ward 318* (CU). **North Carolina.** Camden Co.: Camden, marshy area, 19 Jul 1938, *Godfrey 5326* (GH, US). New Hanover Co.: Wilmington, 10 Jun–6 Jul 1898, *Ashe s.n.* (NY); Wilmington, shore of the NE Cape Fear River, at bridge of US 74-76-421, 10 Oct 1966, *Bradley & Sears 3577* (TEX); tidal marsh along Cape Fear at Wilmington Docks, 13 Sep 1949, *Radford 5040* (NY); Wilmington, 28 Sep 1925, *Uhler 95* (US). Pasquotank Co.: Disturbed swamp margin, 0.7 mi W of Weeksville on NC 170, 31 Jul 1958, *Ahles & Duke 48144* (GH). Perquimans Co.: Marsh on Perquimans River 3.6 mi SE of New Hope, 26 Sep 1967, *Bozeman & Radford 11538* (TEX). Vance Co.: Swampy, thicketed shores of mill pond on Weaver Creek 3.7 mi N-NE of Epsom, 1 Oct 1956, *Ahles & Leisner 20413* (MICH). Washington Co.: Plymouth, wet place, 18 Oct 1958, *Radford 42444* (NY). **Ohio.** Champaign Co.: Mad River, 28 Aug 1838, *Samples s.n.* (MICH). Erie Co.: Florence, Bristol's Woods, 1 Aug 1897, *Moseley s.n.* (US). Hamilton Co.: Madeira, shaded swamps, 30 Aug 1910, *Braun s.n.* (US). Lorain Co.: Oberlin, South Woods, 6 Aug 1894, *Dick s.n.* (MICH); Oberlin, South Woods, 6 Aug 1894, *Ricksecker s.n.* (CU, US). Mahaming Co.: Youngstown, 4 Sep 1919, *Young s.n.* (CU). **Pennsylvania.** Adams Co.: Gettysburg, 4 Jul 1887, *Atwood s.n.* (CU). Bucks Co.: Yardley, 8 Aug 1925, *Dreisbach 3713* (MICH); Yardley, swampy woods, 27 Sep 1924, *Long 31511* (MAK). Chester Co.: West Chester, Marshall's Woods, 5 Sep 1927, *Edmondson 6384* (NY); Westtown, 28–31 Aug 1923, *Pennell 11838* (NY). Dauphin Co.: Harrisburg, Aug 1887, *Small s.n.* (US). Lancaster Co.: 1.5 mi N of Union, 2 Sep 1953, *Schaeffer 45270* (US). Lehigh Co.: Vic. Allentown, Mountainville, 4 Aug 1900, *Dowell 950* (US); Bethlehem, Aug 1832, *Moser s.n.* (NY). Philadelphia Co.: Without specific locality, in 1855, *Garrigues s.n.* (MICH). Wayne Co.: NE of Lakewood, bog, 3 Jul 1957?, *Dix 1847* (CU). **Rhode Island.** Newport Co.: Jamestown, Narragansett Bay, 24 Aug 1936, *Bailey s.n.* (BH). Providence Co.: Providence, Jul 1887, *Olney s.n.*

(CU). SOUTH CAROLINA. Beaufort Co.: Cat-tail swamp, 0.5 mi W of Co. 111 on Co. 33, 6 Sep 1956, *Ahles & Bell 18009* (UNC). Colleton Co.: Ashepoo River, 8 Sep 1915, *Alexander 228* (US); low woods, 0.4 mi NE of Co. Hwy. 40 on Pon Pon Chapel Rd., N of Jacksonboro, 5 Sep 1956, *Bell 4680* (UNC). Georgetown Co.: Pee Dee Game Mangt. Area, 5 Oct 1974, *Culp 95* (CLEMS); abandoned rice fields on Dirlton Plantation, 5 Oct 1974, *Melvin 237* (CLEMS). Marlboro Co.: Naked Creek, E of Wallace nr. SC 9, fresh-water marsh, 30 Sep 1956, *Radford 9110* (DAO, UNC). TENNESSEE. Henry Co.: Swamp along flood plain of Terrapin Creek, ca. 0.5 mi NE of confluence of Terrapin and Sugar Creek, ca. 2.5 mi NW of Jones Mill, 11 Oct 1981, *Webb 4359* (NY, TENN); swamp on the farm of Robert Orr, about 3 mi NW of Jones Mill, 31 Aug 1975, *Webb & Hainley s.n.* (TENN, VDB). Johnson Co.: Shady Valley, 21 Aug 1951, *Barclay 56* (TENN); Shady Valley, boggy woods, 21 Jul 1948, *Fairchild et al. 11701* (TENN); Shady Valley Swamp, 22 Aug 1933, *J.K.U. & Sharp 346* (TENN); Cole's Bog nr. Shady Valley, 13 Oct 1940, *Sharp 1117* (TENN). VERMONT. Addison Co.: Leicester, swale nr. Mud Pond, 28 Aug 1922, *Dutton s.n.* (CU). Windham Co.: Vernon, 1 Aug 1895, *Grout s.n.* (US). VIRGINIA. Alexandria: Nr. hunting ground, 19 Aug 1902, *Shull 229* (NY). Bedford Co.: Without specific locality, Sep 1867, *Curtis s.n.* (CU). Charles City Co.: Weyanoke, fresh tidal marsh along Kittewan Creek, 18 Sep 1939, *Fernald & Long 11332* (GH). Henrico Co.: In an edge of open swamp along Interstate 95S, ca. 6 mi N of Richmond, growing with *Polygonum sagittatum,* 14 Aug 1982, *Park 351* (BH). James City Co.: Tidal marsh along Chickahominy River about 5 mi W of Toano, 13 Aug 1939, *Menzel 307* (GH, MICH). New Kent Co.: Border of fresh tidal marsh by Chikahominy River, SE of Windsor Shade (Boulevard Post Office), 9 Sep 1941, *Fernald & Long 13600* (GH); Chickahominy River at Lanexa, 30 Jul 1921, *Grimes 4124* (NY). Norfolk Co.: Fresh tidal reed-marsh along Northwest River below NW Norfolk Co., 14 Sep 1945, *Fernald & Long 14932* (GH, US). Princess Anne Co.: Blackish marsh of Blackwater River, Pungo Ferry, 22 Sep 1933, *Fernald & Griscom s.n.* (GH); fresh tidal reed-marsh along Blackwater Creek, below Blackwater, 18 Sep 1945, *Fernald & Long 14933* (GH, NY); without specific locality, 3–19 Sep 1905, *Mackenzie 1687* (NY). Washington Co.: Bristol, 12 Sep 1897, *Pringle s.n.* (US). WASHINGTON. Klickitat Co.: Husum, Mt. Paddo, Falcon Valley, Aug 1904, *Suksdorf s.n.* (L). WEST VIRGINIA. Hampshire Co.: Vanderlip, in springy place, 4 Sep 1946, *Davis & Davis 7817* (UNC). Kanawha Co.: Kanawha State Park, 9 Aug 1966, *Denninson s.n.* (NLU). Raleigh Co.: Glen Daniel, 15 Jul 1931, *Col.?* (MIN). WISCONSIN. Jackson Co.: Wooded shore of Sand Creek Pond, Melrose Township, 13 Aug 1956, *Hartley 2268* (NY); boggy woods bordering Ketchum Creek, 3 Sep 1958, *Hartley & Thorne 6079* (GH). Lincoln Co.: Corning, along a ditch, 30 Aug 1955, *Seymour & Jones 16388* (GH). Monroe Co.: Marshy shore of Angelo Lake, 2 Sep 1956, *Hartley 2829* (US).

7. **Polygonum stelligerum** Cham., Linnaea **8:** 131. 1833. *Tracaulon stelligerum* (Cham.) Greene, Leafl. Bot. Observ. Crit. **1:** 22. 1904. *Truellum stelligerum* (Cham.) Sojak, Preslia **46:** 149. 1974. Type. Brazil. Without specific locality and date, *Sellow s.n.* (one sheet believed to be an isotype at P!).

Polygonum acanthophyllum Lindau, Beibl. Bot. Jahrb. **48:** 12. 1894. Type. Argentina. "Buenos Ayres, ad litora maris," without date, *Bettfreund 127* (lectotype, B, designated by Park, Brittonia **38:** 399. 1986; photo of lectotype, F!). Syntype. Argentina. "Entrerios, Rio Cupalen, in inundatis," *Lindau 262* (not seen).

Erect to scandent *herb,* 0.7–1.5 m tall; stem stout, branched, angled, green, moderately to densely covered with 4- to 8-rayed sessile stellate hairs, hispid with bristlelike simple multiseriate hairs to 3 mm long at least in upper portion of internodes; angles of the stem usually armed with bristlelike recurved prickles to 2 mm long. *Leaves* long-petioled; blades narrowly triangular, 9.5–20 cm long, 5.5–14 cm wide, acuminate at the apex, deeply sagittate to weakly hastate with acute triangular lobes at the base; both surfaces moderately to densely pubescent with 4- to 6-rayed sessile stellate hairs and with simple multiseriate hairs to 2 mm long; margins ciliate, often prickly toward the base; midveins and major veins beneath densely stellate-pubescent, hispid, bearing recurved prickles; petioles 4–10 cm long, densely stellate-pubescent, hispid, retrorsely prickly. *Ocreae* scarious, cylindrical, 2–4.5 cm long, oblique at the apex, densely stellate-pubescent and hispid with bristlelike hairs to 3 mm long. *Inflorescences* terminal or in the axils of upper leaves, elongate, 10–25 cm long, branched 1–3 times, often geminate, densely stellate-pubescent throughout, glandular-pubescent in upper portion, hispid below; glandular hairs stalked, ca. 0.5 mm long; lower inflorescence bracts resembling ocreae, 0.7–1 cm long; upper inflorescence bracts reduced, lacerate, hispid, 3–8 mm long; each inflorescence branch terminated by a dense headlike cluster of many flower fascicles. *Flower fascicles* 2- to 3-flowered, closely spaced; bracts elliptic, hispid, ca. 4 mm long; bracteoles broadly elliptic, densely stellate-pubescent, often hispid along the midveins, ca. 4 mm long; pedicels short, 2–3.5 mm long, glabrous, almost included in bracteoles; perianth 4-parted, 3–4 mm long, white; tepals broadly elliptic to ovate, not becoming fleshy; stamens 6(–8), inserted at the base of perianth in two whorls; inner stamens two, ca.

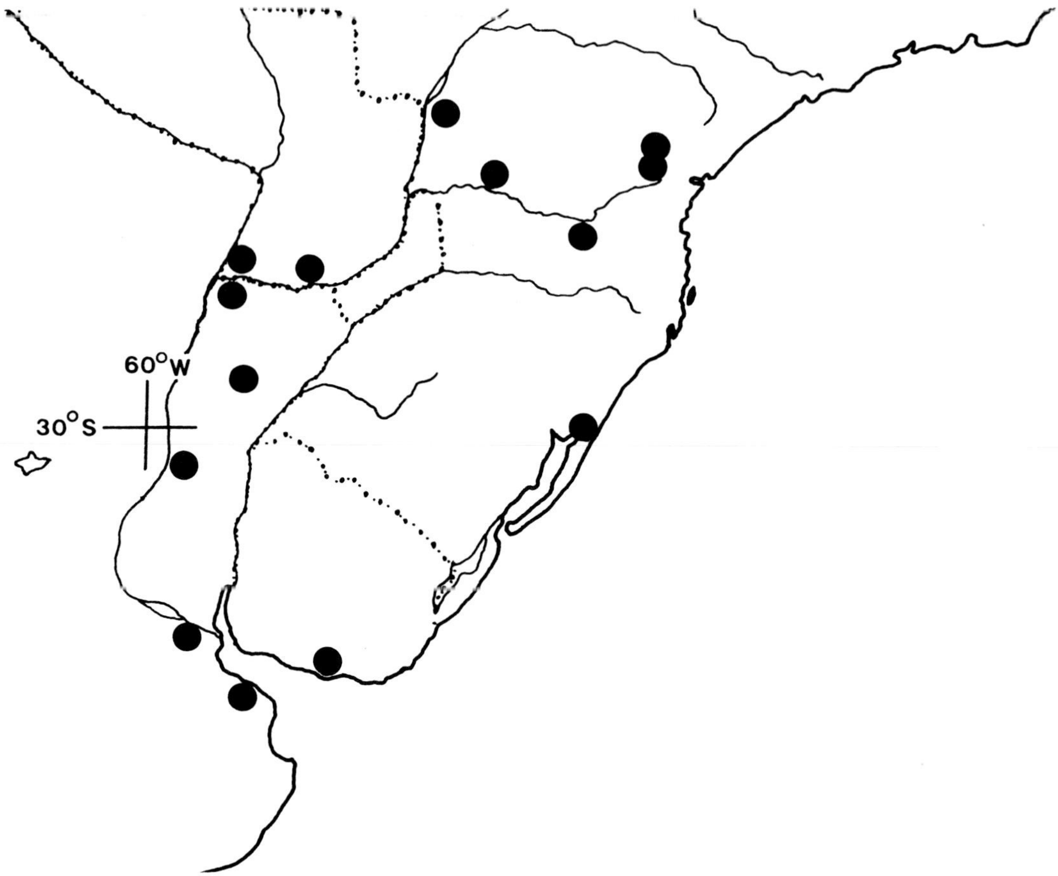

Fig. 19. Distribution of *P. stelligerum.*

2 mm long; outer stamens usually four, ca. 1.3 mm long; style one, deeply 2-cleft, 1.5–2 mm long, included; stigmas two. *Achenes* biconvex, obovoid in outline, 3–4 mm long, dark brown to black at maturity, smooth, shiny.

Distribution. In southeastern South America, from southern Brazil, through eastern Paraguay and Argentina, to Uruguay (Fig. 19).

Polygonum stelligerum is endemic to South America, with the center of distribution in southern Brazil. It is morphologically distinct from other South American taxa, and no close relatives could be found elsewhere. A distant relationship with *P. arifolium* of eastern North America might be suggested on the basis of overall similarities in leaf shape and some reproductive characters, but the two species are very different in many other respects. In *P. stelligerum,* the inflorescence branch is terminated by a dense headlike cluster of many flower fascicles, and stems are very stout and densely stellate-pubescent.

Additional specimens examined. ARGENTINA. BUENOS AIRES: Delta del Paraná, canal Onday Paraná de Las Palmas, 12 Apr 1956, *Burkart 20010* (GH); Punta Lara, 28 Apr 1931, *Cabrera 1721* (F); Río Santiago, 7 May 1939, *Cabrera 5240* (F); Punta Lara, 27 May 1945, *Krapovickas 2281* (US). CORRIENTES: Dep. San Martín, Colonia c. Pellegrini, 22 Feb 1976, *Krapovickas 29520* (IPCN); Dep. Itati, Estancia, 23 Apr 1960, *Pedersen 5563* (A, MO); Dep. Mercedes, cnia. Pellegrini, antes del Piedraplen, 6 Mar 1973, *Quarin & Schinini 1066* (MO, P). ENTRE RÍOS: Dep. La Paz, Isla Alcaraz nr. Piedras Blancas, open woodland on low, floodable ground, 4 Jul 1967, *Pedersen 8302* (A, K).

BRAZIL. PARANÁ: Piraquara–Paraná (Rio Iarai), Jan 1967, *Dombrowski* (*2361*) & *Kuniyoshi* (*2103*) (A); Mun. Guaira, Parque Nacional Sete Quedas, 200 m, 6 Apr 1961, *Hatschbach 7986* (L, US); Castro Rio Iapó, 985 m, 14 Apr 1962, *Hatschbach 9101* (L); Mun. Guaira, Parque Nac. 7 Quedas, 24 Mar 1977, *Hatschbach 39803* (BH, MO, UEC); Mun. Mandirituba, Rio Mauricio, 23 Feb 1978, *Hatschbach 41454* (BH, NY, UEC); Mun. Xambré, Pôrto Byington, 23 Aug 1967, *Hatschbach & Haas 15858* (L); Mun. Pôrto Victoria, Rio Iguaçu, Salto Grande, 14 Feb 1966, *Hatschbach et al. 13888* (P, US); Serrinha, locis hummidis, 4 Jun 1914, *Jonsson? s.n.* (F, MO, US); vic. Pôrto Byington, 23 Jan 1967, *Lindeman & Haas 4387* (NY). **Rio Grande do Sul:** Gravatai, 18 Apr 1949, *Rambo 41083* (F); Lagoa dos Qudros, prope Osorio, 18 Jan 1951, *Rambo 49753* (NO, US); Pôrto Alegre, May 1889, *Reineck s.n.* (K). **Santa Catarina:** Pôrto União, 750 m, 22 Apr 1962, *Reitz & Klein 12750* (L, NY, US); Pôrto União, 750 m, 1 Jun 1962, *Reitz & Klein 11672* (US).

PARAGUAY. **Misiones:** Santiago, Estero Tyecuá, deep swamp, among tall reeds, 2 Apr 1955, *Pedersen 3261* (MO, US); Guazu, 25 Jan 1944, *Royas 10968* (A).

URUGUAY. **San José:** Arazati, 17 Mar 1957, *Arrillaga 651* (F); Arazati, 16 Oct 1965, *Puerto 5347* (US).

8. **Polygonum muricatum** Meisner, Monogr. Polyg. 74. 1826. *Tracaulon muricatum* (Meisner) Greene, Leafl. Bot. Observ. Crit. **1:** 22. 1904. *Polygonum strigosum* var. *muricatum* (Meisner) Steward, Contr. Gray Herb. **5(88):** 89. 1930. *Persicaria muricata* (Meisner) Nemoto, Fl. Japan, Suppl. 173. 1936. *Truellum muricatum* (Meisner) Sojak, Preslia **46:** 148. 1974. Type. Nepal. Without specific locality, in 1821, *Wallich s.n.* [holotype, G-DC (microfiche at NY!); isotype, NY!)].

Polygonum asperulum Wallich, Wallich's Catalogue no. 1724. 1829, nom. nud. Authentic specimen. Nepal. Without specific locality, in 1829?, *Wallich 1724* [G-DC (microfiche at NY!), K!].

Polygonum hastatosagittatum var. *latifolium* Makino, Bot. Mag. (Tokyo) **17:** 120. 1903. *Polygonum nipponense* Makino, Bot. Mag. (Tokyo) **23:** 89. 1909. *Persicaria nipponensis* (Makino) Gross ex Nakai, Fl. Saishu & Kwan Isl. 41. 1914. *Persicaria nipponica* (Makino) Nakai, Rigakkai **24:** 299. 1926. *Truellum nipponense* (Makino) Sojak, Preslia **46:** 148. 1974. Type. Japan. Kochi: Prov. Tosa, Takaoka-gun, Sakawa, in 1884, *Makino 16082* (lectotype, MAK!, designated by Park, Brittonia **38:** 401. 1986). Syntypes. Japan. Kanagawa: Prov. Sagami, Hakone, 24 Sep 1886, *Makino 16080* (MAK!); Kawasaki, *Makino s.n.* (not seen). Kochi: Prov. Tosa, Arioka in Hata-gori, 4 Nov 1885, *Makino s.n.*(not seen); Kambara, 17 Oct 1892, *Makino 10688* (MAK!); Gohokumura, Edagawa, 22 Oct 1892, *Makino 16087* (MAK!); Godaisan, 29 Sep 1892, *Makino 16084* (MAK!); Koishiki nr. Kochi, Sep 1888, *Makino 16083* (MAK!). Nagasaki: Prov. Tsushima, Kuta, 17 Oct 1902, *Hirata 469* (MAK!, TI!). Saitama: Prov. Musashi near Omiyahachinam, 5 Oct 1879, *Yatabe & Matsumura s.n.* (TI!). Yamaguchi: Prov. Suo, Ouchi-mura, 4 Oct 1891, *Nikai 706* (TI!).

Polygonum thunbergii var. *spicatum* Léveillé, Bull. Soc. Bot. France **51:** 423. 1904. Type. Japan. Aomori: Aomori, Sep 1902, *Faurie 5132* (holotype, E!).

Polygonum oliganthum Diels, Notes Roy. Bot. Gard. Edinburgh **5:** 260. 1912. Type. China. Yunnan: Lichiang Range, 10–11,000 ft, Sep 1906, *Forrest 2985* (holotype, E!).

Polygonum benguetense Merrill, Philipp. J. Sci., C. Botany **10:** 301. 1915. *Polygonum strigosum* var. *benguetense* (Merrill) Steward, Contr. Gray Herb. **5(88):** 90. 1930. *Truellum benguetense* (Merrill) Sojak, Preslia **46:** 145. 1974. Type. Philippines. Benguet: Baguio, May 1914, *Merrill 9682* (lectotype, NY!, designated by Park, Brittonia **38:** 401. 1986). Syntypes. Philippines. Benguet: *Merrill 768, Ramos 13483,* and *Robinson 14120* (none seen).

Scandent to ascending *herb* arising from decumbent radicant base, 0.3–1.2 m tall; stem green to light green, often branched extensively at the base, angled, armed with recurved prickles at least in upper portion of the internodes; prickles short, stout, ca. 1 mm long. *Leaves* petioled; blades thin, ovate to elliptic-oblong, 2.5–6.7 cm long, 1.5–3 cm wide, acute to acuminate at the apex, truncate to weakly cordate at the base, antrorsely ciliate at the margin; upper surfaces green, almost glabrous to sparsely pubescent with short appressed simple multiseriate hairs toward the apex, rarely bearing minute 2- to 5-rayed tufted hairs; lower surfaces light green, almost glabrous, often bearing a few minute 4- to 6-rayed sessile stellate hairs; midveins beneath usually antrorsely prickly with short broad-based prickles; petioles relatively long, 0.8–2 cm long, retrorsely prickly, rarely unarmed. *Ocreae* scarious, cylindrical, 1–2 cm long, glabrous, truncate and long-ciliate at the apex. *Inflorescences* terminal or in the axils of upper leaves, elongate, 4–18 cm long, branched 1–5 times, densely pubescent with minute simple unicellular hairs and glandular hairs; glandular hairs stalked, to 0.5 mm long; branches of inflorescence often slender, somewhat flaccid; inflorescence bracts resembling ocreae but somewhat reduced, 2–3 mm long, ciliate at the margin; each inflorescence branch terminated by an interrupted panicle or a headlike cluster of 6–15 flower fascicles. *Flower fascicles* 2- to 3-flowered; bracts broadly elliptic to ovate, 2–2.5 mm long, acuminate at the apex, ciliate at the tip, usually glabrous but lower ones often bearing a few glandular hairs; bracteoles elliptic, 2–2.5 mm long, glabrous; pedicels short, 2–3 mm long, glabrous,

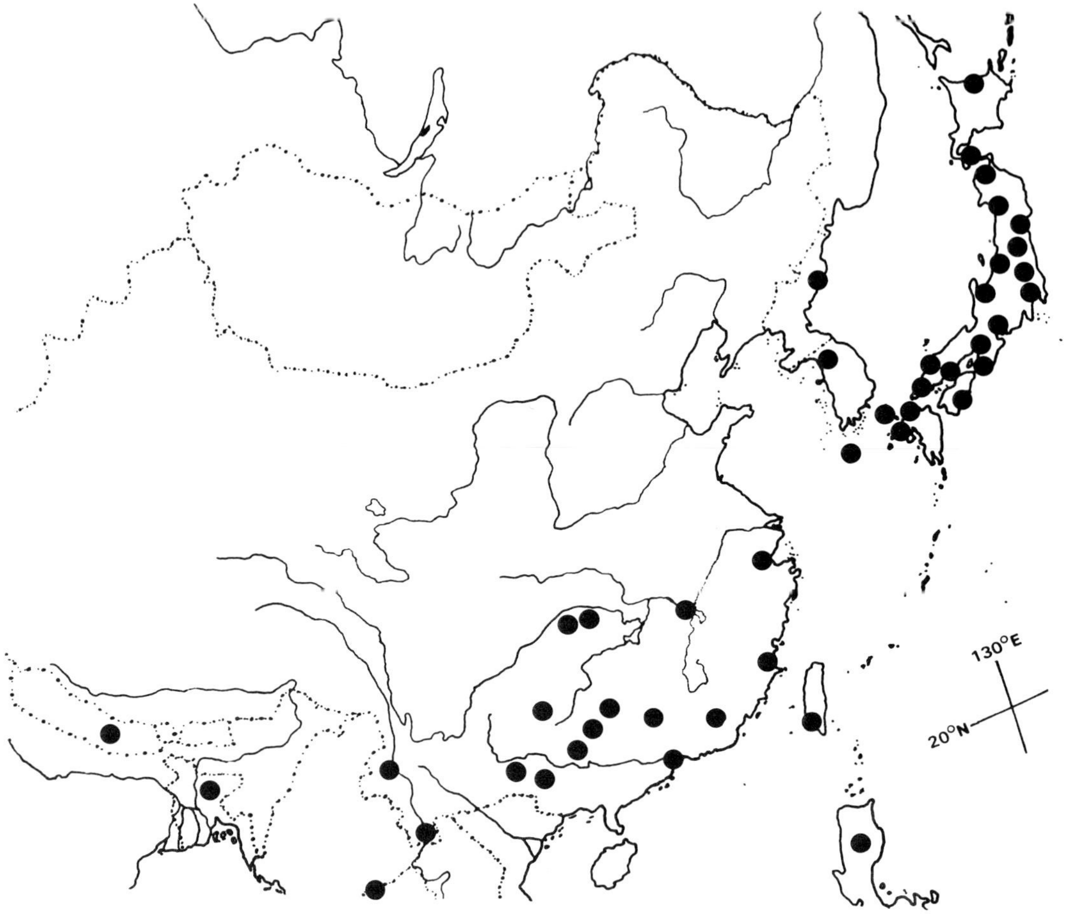

Fig. 20. Distribution of *P. muricatum.*

included in bracteoles; perianth 5-parted, greenish-white with purple tinge at the tip, 2–3 mm long; tepals broadly elliptic, obtuse, not becoming fleshy; stamens 5–9, commonly 6–7, inserted at the base of perianth in two whorls; inner stamens three, ca. 1.5 mm long; outer stamens usually 3–4, ca. 1 mm long; style one, deeply 3-cleft, ca. 1 mm long, included; stigmas three. *Achenes* brown, trigonous, ovoid in outline, acute at the apex, smooth, shiny, 1.5–3 mm long.

Distribution. In moist situations, in partial shade; occurring naturally in wet places in lowlands and along streams; from Japan, through Korea and China, to northern India; common in Japan, Korea, and China (Fig. 20).

Polygonum muricatum shows extreme variation in its inflorescence. Plants of southern China and the Assam region tend to have somewhat slender, flaccid, loose inflorescences as compared to those of eastern China and Japan. In the latter regions, a high proportion of collections has relatively stout, ascending inflorescences with many flower fascicles aggregated into headlike clusters.

The type specimen cited by Meisner (1826) in his original description was collected from Nepal, the western extreme of its range. As in many other plants at the edge of their distribution range, Meisner's type does not well represent most of the plants in the center of the distribution range; it appears to be a western extreme form of *P. muricatum.* The other extreme form, eastern race, has been recognized by Makino (1909) as a distinct species, *P. nipponense.* Makino pointed out that it is distinct from *P. muricatum* by its dense, capitate inflorescence and completely enclosed achene. On the basis of these descriptions, the plants from northern India and China have been referred to *P. muricatum,* while Japanese plants were all identified as *P. nipponense.* The difference in inflorescence between these two extreme forms is of a magnitude that might justify recognition of distinct taxonomic entities. However, there is no geographic partitioning of variation in this character, and broad geographic intergradation can be recognized as in *P. hastatosagittatum.* In addition, *P. muricatum* has completely enclosed achenes; the segregation of *P. nipponense* from *P. muricatum* by Makino (1909) based on achene character is also unwarranted. Therefore, *P. nipponense* can hardly be considered as distinct from *P. muricatum,* and it seems necessary to treat *P. nipponense* as a synonym of *P. muricatum.*

Additional specimens examined. CHINA. ANHWEI: Wu Yuan?, 18 Aug 1825, *Ching 8907* (K); without specific locality, 2 Oct 1951, *Col.?* (NSM). CHEKIANG: Hangchow, 200 ft, in 1921, *Allison 28* (GH). FUKIEN: Minhow Hsien, Pehling, growing by side of stream, 6 Aug 1923, *Chung 2158* (K); Foochow and vic., hillsides nr. Fukien Christian Univ., 1 Oct 1924, *Metcalf et al. 566* (MT). HUNAN: Sinning Hsien, Ma-Ling-Tung, 23 Sep 1935, *Fan & Li 538* (GH, L). HUPEH: *Metasequoia* region of Lichuan Hsien, vic. of Zhuanjiaowan on the E side of the valley, 1500 m, 7 Oct 1980, *Bartholomew et al. 2013* (NY); Chienshih, without date, *Henry s.n.* (K). KIANGSI: Lushan, in 1975, *Cheng 143* (E). KWANGSI: Langshian, by open stream side, 1200 ft, 4 Jun 1928, *Ching 5671* (NY); Seh-feng Dar Shan, S Nanning, 25 Oct 1928, *Ching 8207* (NY); Lin Yuin Hsien, 1000 m, 11 Jun 1933, *Steward & Cheo 643* (NY); Yung Hsien, Ta Tse Tsuen, valley shade, 380 m, in 1933, *Steward & Cheo 755* (GH, NY). KWANGTUNG: Tinghushan, 60 m, in 1978, *Chow et al. 78047* (E, NY, UNC); Loh Fau Mt., 21 Aug 1917, *Levine 1539* (MO); Loh Fau Mt., 9–27 Aug 1917, *Merrill 11025* (NY); Lingnan, Iu Village and Yeung Uk Village, 13 Jun & 7 Jul 1924, *To & Tsang 12467, 12694* (NY, P). KWEICHO: Pin-fa, in 1905, *Cavalerie s.n.* (K). YUNNAN: Mt. Tali Range, Lat. 25°40′N, 6500–9000 ft, Jul 1910, *Forrest 6932* (E, K); Szemao, without date, *Henry 11918* (K, L, LE, MO, NY, US).

HONG KONG. Without specific locality, 8 Nov 1893, *Bodinier 499* (P); without specific locality, colonized in a blocked irrigation ditch, 29 Oct 1972, *Woo & Woo 776* (MICH).

INDIA. ASSAM: Khasi Hills, Pynursla, marsh, 4000 ft, 20 Aug 1949, *Chand 2048* (L); Khasi Hills, Mauplang, along the river in gravel, 6000 ft, 18 May 1954, *Chand 7697* (L); Mairang, 4500 ft, 31 Oct 1871, *Clarke 16164B* (K); Shillong, 5000 ft, 23 Oct 1872, *Clarke 17683A* (K); Khasi Hills, Surareen, 5000 ft, 12 Sep 1885, *Clarke 40419C* (NY); Mt. Khasia, 5–6000 ft, without date, *Hook. f. & Thomson s.n.* (GH, P); Khasi Mts., without date, *Hook. f. & Thomson s.n.* (K); Cherrapunjee, marsh, 4000 ft, 30 Apr 1952, *Koelz 29577* (K); Nongkhlaw, 21 Sep 1956, *Panigrahi 3319* (L); Khasi and Jaintia Hills, area in front of the Nirada Bungalow, 23 Nov 1956, *Panigrahi 3792* (L); Deputasion?, in 1835, *Wallich s.n.* (NY).

JAPAN. AICHI: Nagoya-shi, 23 Oct 1910, *Amano s.n.* (MAK 16170). AKITA: Prov. Ugo, Akita-shi, Nibetsu, ca. 100 m, 28 Oct 1980, *Endo 221* (TUS). AOMORI: Prov. Mutsu, Usoriyamako in Shimokata Peninsula, 3 Sep 1964, *Yoshioka & Sugawara s.n.* (TUS). FUKUOKA: Kasuya Co., Tatara, sea level, 4 Oct 1926, *Ichigawa 126* (US). HOKKAIDO: "Circa Hakodate," 1861–1863, *Albrecht s.n.* (LE); Hakodate, 25 Sep 1886, *Faurie 1417* (P); Hakodate, 26 Sep 1886, *Faurie 1459* (K); Hakodate, 27 Sep 1888, *Faurie 3329* (K); Abashiri, Mokoto-ku, 8 Sep 1974, *Hara et al. s.n.* (TI, TUS). HYOGO: Mineyama-kuogen, Kamioda, Ookouchi-cho, Kanzaki-gun, 940–1040 m, 23 Sep 1983, *Kurosaki 13935* (TI); Minomo in Settsu, 23 Oct 1955, *Togashi 1247* (BH, BM, K, MAK, MICH, MT, NY, P, TUS, US). IWATE: Prov. Rikuchu, Migusawa-machi, 24 Sep 1931, *Hatsuro 129* (TUS). KANAGAWA: Yokoska, without date, *Savatier s.n.* (P). KYOTO: Yodo, Yamashiro, 21 Nov 1952, *Togashi s.n.* (TI). MIYAGI: Shiroishi City, Saikawa, Takijiri, without date, *Hoya s.n.* (TUS); Rikuzen, Sendai, Aobayama Bot. Garden, 21 Sep 1962, *Kiyohiro s.n.* (TUS); Kami-gun, Nakaniida-machi, Naruse River, 23 Sep 1978, *Naito & Ishigawa s.n.* (TUS); Rikuzen, Shiogama, 13 Sep 1959, *Numabe s.n.* (TUS); Rikuzen, Masuzawa, Daiwa-cho, 9 Oct 1972, *Ogura 216* (TUS); Rikuzen, Kaminohara, 10 Oct 1962, *Ohashi s.n.* (TUS). NAGANO: Higashisonogi-gun, 7 Oct 1905, *Chiba s.n.* (MAK 16085); Ookamura, 10 Sep 1958, *Minemura s.n.* (MAK 8682). NAGASAKI: Nagasaki, Sep 1863, *Maximowicz s.n.* (LE, P). NIIGATA: Mt. Koma in Echigo, 4–6 Oct 1950, *Ohwi s.n.* (BH). OKAYAMA: Mitsu-gun, Mitsu-machi, Masera, 50–100 m, 30 Sep 1979, *Ohba & Akiyama 661* (TI). SAITAMA: Urbe Urawa in Musashi, 1 Nov 1950, *Koyama 166* (BH, E, K, L, MICH, MT, NY, US); Sekido in Musashi, 24 Oct 1937, *Okuyama s.n.* (BH); Toda, 26 Sep 1937, *Togashi s.n.* (TI); Musashi, Kami-igusa, 22 Oct 1893, *Watanabe s.n.* (GH, US). SHIMANE: Kakinoki-cho, 18 Aug 1961, *Miyamoto s.n.* (MAK 28585). TOCHIGI: Nikko, Odajiro-ga-hara, 9 Sep 1975, *Murata & Ohashi s.n.* (TI); Nikko-shi, 29 Aug 1910, *Watanabe s.n.* (MAK 16153). TOKYO: Edogawa-ku, 2 Oct 1904, *Makino 16187* (MAK); Kamiitabashi, 29 Oct 1904, *Makino 16191* (E, LE); Edogawa-ku, Komatsukawa, 10 Oct 1908, *Makino 16086* (MAK); Itabashi-ku, Shimura, Nov 1909, *Makino 16156* (MAK); Shimura, in 1924, *Makino 16190* (E, L). TOYAMA: Within the town limits of Yatsuo, 17 km S of the City of Toyama, along damp roadside, 16 Oct 1954, *Kirino 431* (US). WAKAYAMA: Prov. Kii, Kamitondamachi, Nakaoka, 16 Oct 1957, *Sugaya & Toda s.n.* (TUS). YAMAGATA: Yonesawa, 14 Oct 1899, *Nakamura s.n.* (MAK 16145). YAMAGUCHI: Nagato-koku, Abu-gun, Ato-cho, Tokusa, 27 Sep 1969, *Nagato s.n.* (TUS); Kuma-gun, Miwacho, Tateiwa, 13 Oct 1956, *Wada s.n.* (MAK 100382).

KOREA. CHEJU: Cheju Isl., 800 m, Aug 1911, *Taquet 5902* (TI). HAMNAM: Pujeon Plateau, 15 Jul 1934, *To 1570* (SNU); Wonpoong-Doan, 14 Aug 1943, *To*

1587 (SNU), Pujeon Plateau, in 1940, *Col.?* (MICH). **Kyunggi:** Kum-kok, in wet place along the ditch nr. paddy field, 3 Aug 1981, *Park 211* (BH). **Seoul:** Chungryangri, 10 Sep 1934, *Col.?* (MICH).

MACAO. Without specific locality, in 1844, *Callery 68* (P).

NEPAL. Without specific locality and date, *Wallich s.n.* (MO).

TAIWAN. Without specific locality, in 1864, *Oldham 435* (K). **Pingtung:** Nanjenshan, 21 Apr 1978, *Koyama & Kao 8920* (NY).

THAILAND. Prov. Chaiyaphum, Tunkamang, 850 m, 29 May 1974, *Geesink et al. 7065* (K); Bo Luang Plateau, small stream, 600 m, 24 Sep 1958, *Larsen 5231* (A).

9. Polygonum hastatosagittatum Makino, Bot. Mag. (Tokyo) **17:** 119. 1903. *Persicaria hastatosagittata* (Makino) Nakai ex Mori, Enum. pl. Cor. 132. 1922. *Polygonum strigosum* var. *hastatosagittatum* (Makino) Steward, Contr. Gray Herb. **5(88):** 90. 1930. *Truellum hastatosagittatum* (Makino) Sojak, Preslia **46:** 146. 1974. Type. Japan. Chiba: Prov. Shimosa, Ichikawa-shi, Mama, 26 Oct 1890, *Makino 13982* (lectotype, MAK!, designated by Park, Brittonia **38:** 401. 1986). Syntypes. Japan. Chiba: Prov. Shimosa, Ichikawa-shi, Mama, 6 Oct 1895, *Makino 13979* (MAK!); Prov. Kadzusa, Haseya, Sep 1880, *Col.?* (TI!). Hokkaido: Without specific locality and date, *Boehmer s.n.* (not seen). Kochi: Prov. Tosa, Kochi-shi, Nov 1891, *Makino 13977 & 13978* (both MAK!); Kochi-shi, Oct 1892, *Makino 13980* (MAK!); Edagawa-mura, 22 Oct 1892, *Makino 13983* (E!, MAK!); Kochi-shi, Asakura-gun, 23 Sep 1892, *Makino 13891* (MAK!). Taiwan. Taipei, 12 Nov 1896, *Makino 138* (TI!); Taipei, 21 Dec 1897, *Owatari s.n.* (TI!).

Polygonum sagittatum var. *ussuriense* Regel, Mém. Acad. Imp. Sci. Saint Pétersbourg, sér. 7. **4(4):**126. 1861. *Polygonum ussuriense* (Regel) Nakai *in* Matsumura, Icon. Pl. Koisikavenses **4:** 49, pl. 237. 1919. *Persicaria ussuriensis* (Regel) Nakai ex Mori, Enum. pl. Cor. 134. 1922. Type. U.S.S.R. "Am Sungatsch und Kenga-See im Juli und August," *Regel*? [not seen; see Nakai *in* Matsumura, Icon. pl. Koisikavenses **4:** 49. 1919 (*P. korshinskianum* is given as syn.)].

Polygonum cavaleriei Léveillé, Repert. Spec. Nov. Regni Veg. **8:** 172. 1910. Type. China. Guizhou (Kweichow): Kweichow: Pin-fa, 8 Oct 1905, *Cavalerie 2550* (holotype, E!).

Polygonum korshinskianum Nakai, J. Coll. Sci. Imp. Univ. Tokyo **31:** 169. 1911. *Truellum korshinskianum* (Nakai) Sojak, Preslia **46:** 146. 1974. Type. Korea. "Prov. Kenge, fluvium Jalu, Vallis Andori," 6 Sep 1897, *Komarov s.n.* (holotype, TI!).

Erect to scandent *herb* arising from a fibrous root system, 0.3–1.5 m tall; stem green to light green, branched, angled, often producing roots at lower nodes, sparsely armed with recurved prickles along the angles; prickles short, stout, ca. 1 mm long in upper portion of the internodes. *Leaves* petioled; blades usually lanceolate to narrowly elliptic, but often ovate-oblong, 2.5–9.5 cm long, 1–3 cm wide, subacuminate to acuminate at the apex, usually subsagittate to shallowly cordate with small rounded or triangular lobes at the base, often obtuse to truncate, antrorsely ciliate on upper margin, retrorsely ciliate on lower margin; upper surfaces glabrous to moderately pubescent with 3- to 4-rayed minute tufted hairs, usually bearing appressed bristlelike simple multiseriate hairs toward the apex; lower surfaces glabrous to moderately pubescent with 4- to 6-rayed sessile stellate hairs; midveins beneath usually retrorsely prickly with short broad-based prickles, often bearing some 3- to 4-rayed tufted hairs; petioles relatively long, 0.7–5.5 cm long, usually retrorsely prickly. *Ocreae* scarious, cylindrical, 1.5–3.2 cm long, truncate, glabrous, long-ciliate. *Inflorescences* terminal or in the axils of upper leaves, elongate, 10–21 cm long, branched 1–3 times, often geminate, densely pubescent with glandular hairs and 2- to 3-rayed tufted hairs to 0.3 mm long in upper portion; glandular hairs stalked, ca. 0.5 mm long; lower inflorescence bracts resembling ocreae, 0.7–1 cm long; upper ones usually reduced, somewhat funnelform, 3–5 mm long, strongly ciliate along the margin; each inflorescence branch terminated by a dense headlike cluster of many flower fascicles. *Flower fascicles* usually 2-flowered, closely spaced; bracts broadly elliptic to ovate, 2.5–3 mm long, ciliate at the margin, usually glabrous, but often densely glandular-pubescent; bracteoles broadly elliptic, 2–3 mm long, glabrous, often bearing a few glandular hairs along the margin; pedicels long, 4–6 mm long, longer than bracteoles, exserted at maturity, usually glandular-pubescent; perianth 5-parted, pink, 4–5 mm long; tepals broadly elliptic, obtuse, not becoming fleshy; stamens eight, rarely seven, inserted at the base of perianth in two whorls; inner stamens three, 1.2–2 mm long; outer stamens 4–5, 1–1.5 mm long; style one, 3-cleft up to middle, 1–2 mm long, included; stigmas three. *Achenes* trigonous, ovoid in outline, 3–4 mm long, dark brown to black, smooth, shiny, acute at the apex.

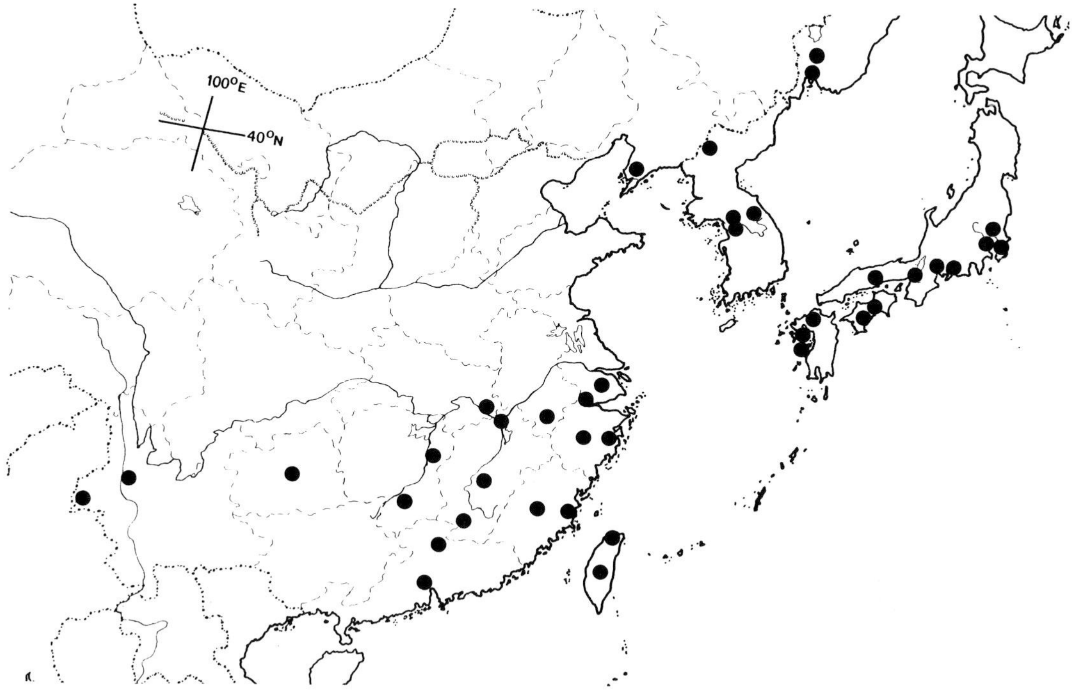

Fig. 21. Distribution of *P. hastatosagittatum.*

Distribution. In moist situations, usually in shade; occurring naturally in wet places in lowlands and along streams; from Japan, through southeastern Siberia, Korea, and Manchuria, to western China (Fig. 21).

Polygonum hastatosagittatum is closely related to *P. muricatum,* but it can be readily distinguished by its long, completely exserted pedicels, which are unique in the section. Chemically, it can be distinguished from *P. muricatum* by two additional flavonoid compounds (Park, 1987).

The most striking variation in *P. hastatosagittatum* is in leaf shape and size. In Figure 22, variation of this character in China is compared with that found in Japan, using measurements taken from herbarium specimens. At the eastern part of its range, which comprises southern Japan, the species displays relatively large, lanceolate to narrowly elliptic leaves with subsagittate to subhastate bases. In China, leaf shape and size become far more variable, from lanceolate to ovate, with the leaf size ranging from 8.5 cm to 2 cm long. A high proportion of collections from the southern extreme of its range in China has relatively small, ovate to broadly elliptic leaves with weakly cordate to truncate bases. They also tend to have somewhat slender, loose inflorescences and extensive decumbent stems. However, these plants do not seem to be consistently distinguishable by any characters other than leaf shape and size, which show some degree of overlap and broad geographic intergradation. Also, plants with similar leaf shape occur sporadically throughout the range of *P. hastatosagittatum*; therefore, no infraspecific taxa are recognized here.

Plants from Manchuria and eastern Siberia have relatively long, linear lanceolate leaves with sharply sagittate bases. In addition, inflorescences are few-flowered and relatively loose. These plants have been recognized as a distinct species (*P. korshinskianum* Nakai) or a variety (*P. sagittatum* var. *ussuriense* Regel). However, it appears to be the poorly developed form of *P. hastatosagittatum,* which is often found in harsh northern environments. The variation may be due to the wide range of habitat types in which this species occurs.

Polygonum hastatosagittatum is frequently confused with *P. dichotomum* (=*P. pedunculare*), and most of the collections from China are determined as that species in herbaria. It can be

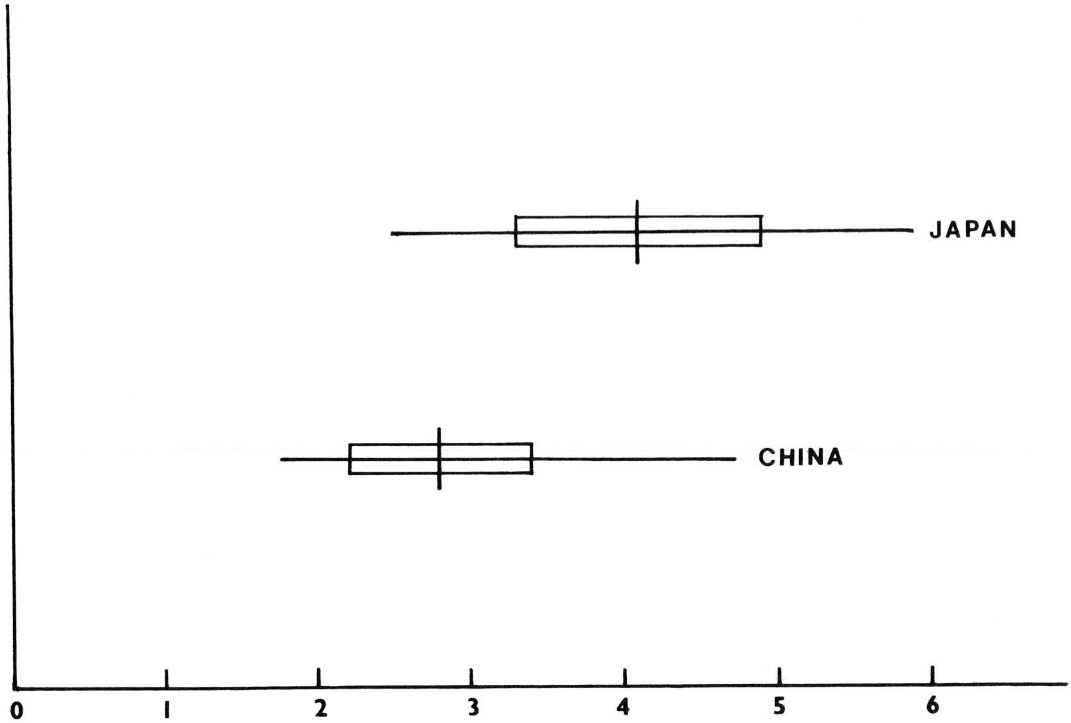

Fig. 22. Variation in leaf length/width ratio of *P. hastatosagittatum* from Japan and China. Measurements were taken from cited herbarium collections. Glyphs show means, standard deviations, and ranges.

easily distinguished from *P. dichotomum* by its long exserted pedicels, three-cleft style, and trigonous achene. In addition, the flavonoid chemistry of *P. hastatosagittatum* is considerably more complex (Park, 1987).

Additional specimens examined. CHINA. ANHWEI: Chien Shan Hsien, Tien Chu Shan, 27 Jun 1936, *Fan & Li 255* (GH, L); without specific locality, 2 Sep 1957, *Fu 808* (NSM); without specific locality, 20 Jun 1959, *Survey team 2246* (NSM). CHEKIANG: Ko Kan Shan, 1 Sep 1926, *Cheo & Wilson 12880* (GH); along streamside nearby Hangchow College, 11 Oct 1928, *Chiao 18883* (NY); Sze-ton, S of Siachu, 150–600 m, 30 May–1 Jun 1924, *Ching 1700* (GH); Linhai, 6 Nov 1935, *Migo s.n.* (NSM); Hiang Lou?, 26 Sep 1876, *Poli s.n.* (P). FUKIEN: Without specific locality and date, *Ching 6774* (NY); Foochow, Pei Ling Mts., 19 Sep 1897, *Carles s.n.* (E); Diongloh and vic., 3 Aug 1926, *Tai 11687* (MT). HUNAN: Changning Hsien, Yang-Shan, 2 Aug 1935, *Fan & Li 392* (GH, L, P); without specific locality, 14 Aug 1948, *Liu 389* (PE); Tshangscha, 11 Oct 1917, *Handel-Mazzetti 2262* (US). HUPEH: Yang Hsin Hsien, 13 Sep 1928, *Choo 18373* (CU, NY); without specific locality and date, *Henry 2793* (GH). KIANGSI: Lu Shan, Hsiu Feng Su, in ditch, 2 Sep 1932,*Cheo 243* (K); without specific locality, in 1868 and 1873, *David s.n.* (P); Ningdu, 21–23 Jul 1921, *Hui 451* (P); Kiennan Dist., Sai Hang Chen nr. Tung Lei Village, 1–11 Sep 1934, *Lau 4354* (US); without specific locality and date, *Survey team 2901* (NSM). KIANGSU: Suchow, Tienping-shan, 22 Sep 1933, *Migo s.n.* (NSM). KWANGTUNG: Without specific locality, 15 Dec 1929, *Armacost 4* (NY); vic. Canton, Lung-yentung, 28 Oct 1929, *Chun 7842* (NY); Canton, Nov 1863, *Hance 1482* (P); Ichang and immediate neighborhood, Feb 1887, *Henry 2247, 2793* (K); Wung-Yuen Dist., Yung-yun City and vic., 26 Nov 1932, *Lau 623* (GH, NY); Wung-Yuen Dist., Wong Chuk I and vic., Tsing Wan Shan, without date, *Lau 2286* (GH); Canton and vic., 13 Nov 1916, *Levine 20* (MO); 16 Dec 1916, *218* (MO); 28 Sep 1917, *1647* (GH, MO); Canton, 17 Oct 1883, *Sampson s.n.* (K). KWEICHO: Houang-Tsao-Pa?, without date, *Cavalerie 5532* (P). LIAONING: Hanka and Manchuria, 1898–1899, *Bohnhof 185C* (E, K, NY); Liaotung Peninsula, nr. Sta. Van Fangoo, 20 Jul 1902, *Litvinov 1694* (GH, NY). YUNNAN: Tali, 27 Sep 1888, *Delavay 3199* (P); Shweli Valley, 6000 ft, Aug 1913, *Forrest 11992* (E); without specific locality and date, *Marie 157* (K).

JAPAN. AICHI: Nishio-shi, Okazaki, Nov 1896, *Makino 13886* (MAK); Nagoya-shi, Ryusengi-yama, 13 Oct 1903, *Yamada s.n.* (MAK 13891). CHIBA: Ichigawa-shi, Mama, 24 Sep 1913, *Makino 13902* (L, US); Mohara-shi, Mohara, Oct 1933, *Makino 13896* (MAK); Yatsumi in Kadzusa, 11 Oct 1936, *Okuyama 9620* (BH); Imbanuma in Shimosa, 15 Oct 1939, *Okuyama s.n.* (BH). FUKUOKA: Hakozaki, Kasuyagun, 2 m, 23 Oct 1928, *Ichikawa 10* (BM, L, TAI). HYOGO: Prov. Settsu, Oct 1932, *Yoshino s.n.* (TI). KOCHI: Kochi, 11 Nov 1893, *Faurie 11942* (P). KYOTO: Yodo, Yamashiro, 21 Nov 1952, *Togashi s.n.* (TI). NARA: Nara-shi, Kasuga-yama, Oct 1921, *Makino 13897* (MAK). OKAYAMA: Okayama-shi, 24 Sep 1913, *Makino 13894* (MAK). SAITAMA: Musashi, Adzusawa, 12 Sep 1893, *Col.?* (US); Musashi, Wada-mura, 30 Oct 1893, *Col.?* (GH). TOCHIGI: Maoka-shi, Nakamura, without date, *Sinozaki s.n.* (MAK 13881). TOKYO: Itabashi-ku, Shimura, Nov 1909, *Makino 13898* (MAK); Nerima-ku, Oizumi, 18 Nov 1936, *Makino 13901* (E, MAK); Tokyo, Koiwa, in 1923, *Makino 13900* (L).

KOREA. KANGWON: Without specific locality, 19 Aug 1930, *Nakai 14678* (TI). KYUNGGI: Kum-kok, wet place nr. abandoned paddy field, 3Aug 1981, *Park 902* (BH); Mt. Chu-kum, along ditch nr. abandoned paddy field, 5 Aug 1981, *Park 905* (BH); Chung-pyung, in wet, shady place along Han River, 12 Aug 1981, *Park 910* (BH).

TAIWAN. Without specific locality, in 1903, *Makino 13880* (MAK); without specific locality, in 1864, *Oldham 435* (GH, P). TAIPEI: Taipei, NTU Campus, 5 Oct 1960, *Chou et al. 7388* (TAI); Yonankuosiao, 29 Nov 1978, *Kuo 10157* (TAI); vic. Taipei, 18 Dec 1930, *Tanaka 469* (P).

U.S.S.R. MARITIME TERRITORY: Vladivostok, 22 Jul 1926, *Melvil s.n.* (NY); Dist. Primorskaya oblast, ad app. Nicolsk-Ussurysk, 27 Jul 1917, *Kiss s.n.* (MICH).

10. Polygonum strigosum R. Br., Prodr. 420. 1810. *Tracaulon strigosum* (R. Br.) Greene, Leafl. Bot. Observ. Crit. **1:** 22. 1904. *Persicaria strigosa* (R. Br.) Nakai, Rigakkai **24:** 299. 1926. *Truellum strigosum* (R. Br.) Sojak, Preslia **46:** 149. 1974. Type. Australia. NSW: Port Jackson, 1802–1805, *Brown s.n.* (lectotype, BM!, designated by Park, Brittonia **38:** 400. 1986; isolectotypes, BM!, K!). Syntype. Australia. Tasmania, *Brown s.n.* (not seen).

Polygonum sagittatum sensu D. Don, Prodr. fl. nepal. 73. 1825; non L., 1753. Type. Nepal. *Wallich s.n.* [not seen; see Meisner, Monogr. Polyg. 65. 1826 (cited as a syn. of *P. sagittatum* var. *indicum* Meisner)].

Polygonum hispidulum Blume, Bijdr. **11:** 535. 1826. Type. Indonesia. Java: "In montosis Javae," *Blume*? (not seen; see Hook. f., Fl. Brit. India **5:** 47. 1886).

Polygonum sagittatum var. *indicum* Meisner, Monogr. Polyg. 65. 1826. Type. Nepal. Without specific locality, in 1821, *Wallich s.n.* [there are five sheets which are believed to be this collection at K, L, MICH, NY, and US, but no sheets bear Meisner's annotation; see Meisner *in* DC., Prodr. **14:** 133. 1856 (cited as a syn. of *P. horridum*)].

Polygonum muricatum Wallich, Wallich's Catalogue no. 1697 (excl. var. *glabratum*). 1829; non Meisner, 1826; nom. nud. Authentic specimen. Nepal. Without specific locality, in 1821, *Wallich 1697* (L!, MICH!, P!, US!).

Polygonum horridum Buch. ex Roxburgh, Fl. ind. ed. 1832. **2:** 291. 1832. *Polygonum strigosum* var. *horridum* (Buch. ex Roxburgh) Kuntze, Revis. gen. pl. **2:** 559. 1891. Type. India. In 1810, *Mr. J. R.* [not seen; see Hook. f., Fl. Brit. India **5:** 47. 1886 (cited as a syn. of *P. strigosum*)].

Polygonum auriculatum Hook. f. & Thomson ex Hook. f., Fl. Brit. India **5:** 47. 1886, pro syn.

Polygonum strigosum var. *rubrifolium* Kuntze, Revis. gen. pl. **2:** 559. 1891. Type. India. Sikkim, 15 Nov 1875, *Kuntze 6598* (holotype, NY!).

Scandent *herb,* 0.5–1.5 m tall; stem branched, angled, usually armed with recurved prickles along the angles; prickles 1–1.5 mm long. *Leaves* subsessile to long-petioled; blades linear lanceolate to elliptic or narrowly triangular, 5–9.8 cm long, 1–3 cm wide, acute to acuminate at the apex, weakly cordate to truncate or subsagittate to weakly hastate with small obtuse triangular lobes at the base, ciliate at the margin; upper surfaces green, glabrous to moderately pubescent with short appressed simple multiseriate hairs toward the apex; lower surfaces usually glabrous but rarely moderately pubescent with 3- to 6-rayed sessile stellate hairs; midveins beneath usually prickly; petioles retrorsely prickly, 0.2–4 cm long. *Ocreae* scarious, cylindrical, 1.5–3 cm long, truncate, ciliate, usually bearing bristles. *Inflorescences* terminal or in the axils of upper leaves, elongate, 6–16 cm long, branched 1–3 times, glabrous to densely pubescent with glandular hairs, often intermixed with minute 2- to 3-rayed tufted hairs; glandular hairs stalked, ca. 0.5 mm long; inflorescence bracts resembling ocreae but much reduced, 5–8 mm long, upper ones often funnelform, ciliate; each inflorescence branch terminated by a dense headlike cluster of many flower fascicles. *Flower fascicles* 2- to 3-flowered, relatively closely spaced; bracts elliptic to ovate, 2–3 mm long, ciliate, often hispid; bracteoles elliptic, glabrous, 2–3 mm long; pedicels short, 2–4 mm long, often glandular-pubescent; perianth 5-parted, whitish-pink to pink; tepals elliptic, obtuse, 2–4 mm long, not becoming fleshy; stamens 5–7, 1–1.5 mm long; style one, 2- or 3-cleft, 1–1.5 mm long, included; stigmas 2–3. *Achenes* biconvex or trigonous, dark brown to black, 3–4 mm long.

Distribution. From Nepal and northern India (Assam) through southeast Asia to eastern Australia (Fig. 23).

Polygonum strigosum is the most widespread and variable species of section *Echinocaulon.*

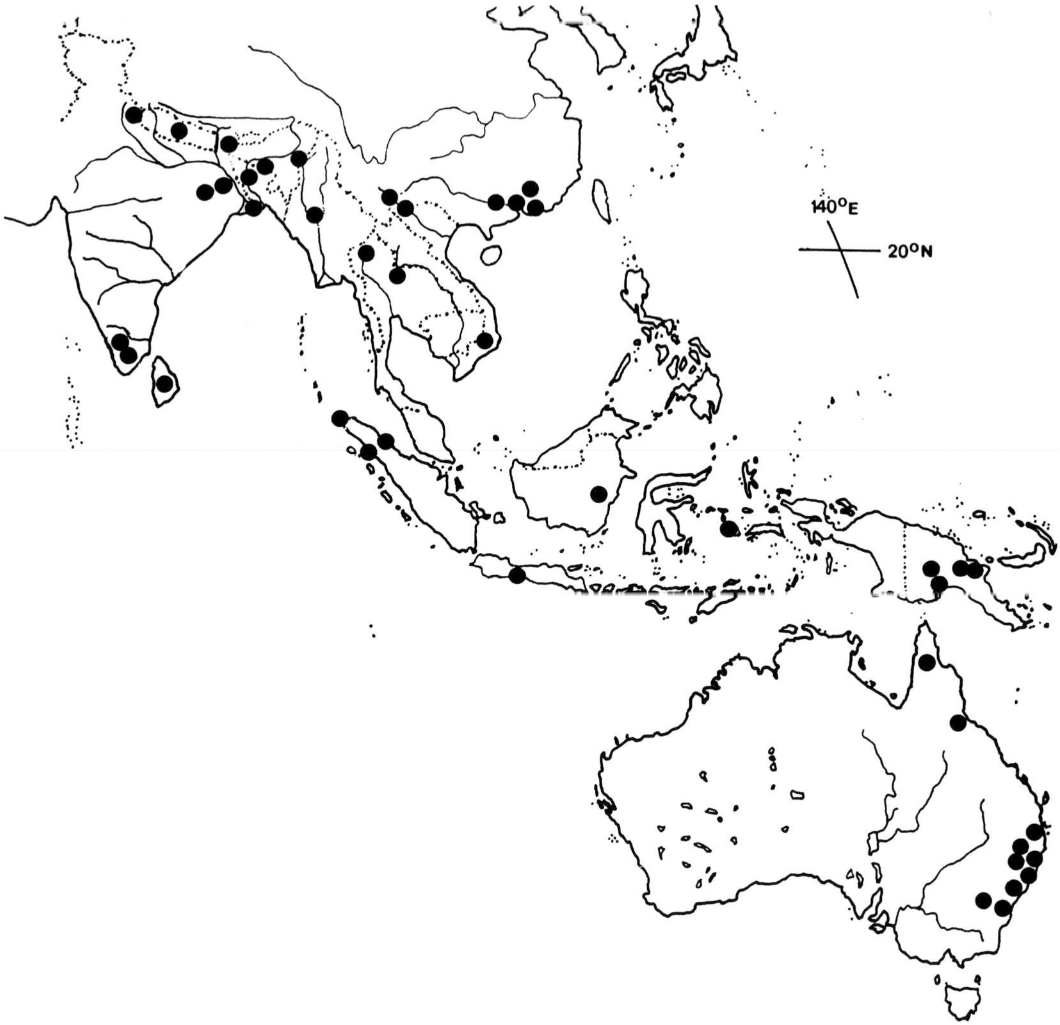

Fig. 23. Distribution of *P. strigosum.*

Extending from northern India through southeast Asia to eastern Australia, it shows remarkable geographical differentiations. Its various forms have received at least five different names, of which the following may deserve recognition.

(1) In northern India (Assam and Nepal), plants have subsessile, strongly ciliate, linear lanceolate leaves with weakly cordate bases. Petioles are very short and almost absent in upper leaves, ranging from 2 to 4 mm long. Ocreae are long-ciliate and glabrous. In addition, they consistently have a three-cleft style and five stamens inserted in one whorl. Achenes are weakly trigonous and rugose. This form has been called *P. horridum* Buch. ex Roxb., but Hooker (1886) treated it as a synonym of *P. strigosum.*

(2) Populations from southern India (Nilgiri and Palni Hills) have short-petioled, narrowly elliptic leaves, with truncate to weakly sagittate bases. In addition, inflorescence branches are ascending and densely pubescent with short two- to three-rayed tufted hairs intermixed with glandular hairs. This form also has been called *P. horridum,* but differs from form (1) in having a two-cleft style and biconvex achene (stamen number, 5–6 in one whorl).

(3) Collections from Sumatra, Java, and Borneo have distinctly petioled, narrowly elliptic to

elliptic leaves with truncate to sagittate bases. Petioles are 1–2 cm long, ranging from ⅕ to $^{3}/_{10}$ of the length of blade. Ocreae are usually bristly at least near the base and long-ciliate. They consistently have a three-cleft style and seven stamens inserted in two whorls. Achenes are trigonous, smooth and shiny. This form nearly approaches *P. muricatum,* but it differs in having bristly ocreae. It has been called *P. hispidulum* Blume.

(4) Plants intermediate between form (1) and (3) occur in southern China (Kwangtung and Yunnan). Their morphology nearly approaches form (1), but differs in having a smooth, shiny, weakly trigonous achene and six stamens inserted in two whorls.

(5) Plants from New Guinea and eastern Australia are less prickly than other forms above and have long-petioled, narrowly triangular leaves with truncate to weakly hastate bases. Petioles are 1–4 cm long, ranging from ⅕ to ⅖ of the length of blade. Inflorescences are somewhat loose, as compared to those of other populations. Styles are more frequently three-cleft than two-cleft, and stamen number varies from five to seven (in two whorls when 6–7). Achenes are smooth, shiny, weakly trigonous or biconvex, and often both types occur on the same plant. Brown's type specimens came from this population. This form is very similar to form (3), but slightly differs in having much longer petioles and narrowly triangular leaves. In addition, ocreae are usually weakly ciliate.

No attempt has been made to provide the formal names for these forms, because forms 2–5 described above are not clearly delimitable entities. The differences among them are primarily in leaf shape and petiole length that would likely have been influenced by habitat differences. Also, intermediate forms occur sporadically throughout the range of distribution, and broad geographic intergradation is recognized. The northern Indian population (form 1) is rather distinct from the rest of populations in having rugose achenes and subsessile, lanceolate leaves; it may deserve at least the varietal status that has been assigned to it by Kuntze (1891). However, it is included here in *P. strigosum* without taxonomic recognition, because of its remarkable similarity in flavonoid chemistry to form 3.

Additional specimens examined. AUSTRALIA. East Australia, in 1865, *Statter s.n.* (BM). NEW SOUTH WALES: Port Jackson District, Mar 1897, *Camfield s.n.* (GH); Waterfall, Feb 1899, *Camfield s.n.* (US); St. Anthony's Creek, Glanmire, ca. 7 mi E from Bathurst, 750 m, 16 Jul 1964, *Constable 5617* (K); Richmond River, Goat Island, 10 mi NE of Woodburn, 23 Nov 1965, *Constable 6451* (K); Mascot, off Wentworth Ave., 31 Mar 1971, *Coveny 3553* (MO); 5 mi NE of Grafton, swampy meadow, 30 Mar 1963, *MacDonald s.n.* (K); Clarence River, in 1869, *Mueller s.n.* (K); Clarence River, in 1871, *Mueller s.n.* (MO); 6 mi S of Ballina, along Pacific Hwy., 17 Jul 1966, *Richards 31648* (L); 12 mi N of Nowra, 9 Feb 1933, *Rodway 1038* (L); 8 km from Wisemans Ferry on lower Portland Rd., 20 Mar 1976, *Wilson 1558* (MO); ca. 4 km NE of Williamtown on Port Stephens Rd., 5 Jul 1979, *Wilson 2409* (K); Emigrant Creek Dam, off Friday Hut Rd., NNW of Ballina, 1 Jan 1980, *Wilson 3099* (K); northern outskirts of Casino, 12 May 1981, *Wilson 3854* (K); 3 km along Diamond Head Road from Mooreland, 18 Dec 1981, *Wilson 3916* (K). QUEENSLAND: Eubenangee Swamp, N of Garradunga, 1 Dec 1941, *Blake 14489* (L); Mt. Fox, Sep–Dec 1949, *Clemens s.n.* (GH, MICH); Brisbane River, Oct 1827, *Cunningham 1000* (BM); Rockingham Bay, in 1869, *Mueller s.n.* (K).

PAPUA NEW GUINEA. Div. Central New Guinea, Komopa, Wissel Lakes, 1755 m, 14 May 1960, *Vink & Schram 8604* (L); Eastern Highlands Dist., on grassy floodbanks of O'awa River, 1400 m, 16 Oct 1959, *Brass 32100* (L); Eastern Highlands Dist., Kainantu Subdist., Aiyura, 5000 ft, 21 Dec 1971, *Henty NGF49176* (L); South Highlands Dist., Tari Subdist., Tigibi, 13 Jun 1966, *Vink 16887* (A).

BANGLADESH. East Bengal, without date, *Griffith 4124* (GH, K); Ganges Plain, without date, *Thomson s.n.* (GH, L); Silhet, without date, *Wallich 1697/2* (L, NY); Silhet, in 1832, *Wallich s.n.* (P).

BURMA. Sagaing Dist., Ava, 27 Mar 1807, *Griffith 4124* (K).

CHINA. KWANGSI: Bako Shan, W. Poseh, 24 Sep 1928, *Ching 7636* (NSM, NY). KWANGTUNG: Canton and vic., 21 Aug 1917, *Levine 1412* (MO, US); Loh Fau Shan, 9–27 Aug 1917, *Merrill 10808* (NY); without specific locality, 4 Jul 1963, *Ting & She 668* (NSM); Suni, without date, *Wang 38403* (TAI); without specific locality, Jan 1964, *Col.?* (PE). YUNNAN: Sjemes?, without date, *Henry 12187* (NY); Ping-pien Hsien, 5 Jul 1934, *Tsai 60667* (A).

HONG KONG. Without specific locality, in 1874, *Forbes 362* (E).

INDIA. "Plants from the Botanical Garden of Calcutta and Serampore," 1834–1841, *Voigt 196, 198 & 146* (GH); without specific locality, in 1821, *Wallich s.n.* (MO, US); without specific locality and date, *Wallich s.n.* (L); "Peninsula Indiae orientalis," without date, *Wight 2456* (GH, K, L, NY). ASSAM: Upper Assam, without date, *Jenkins 505* (GH); Khasi Hills, Mongpoh, 2000 ft, 11 May 1949, *Koelz 22665* (L, MICH); Khasi Hills, Mongpoh, 2000 ft, 7 Jul 1949, *Koelz 23072* (L, MICH); Orang, Mar 1902, *Mokim 22* (P); Margherita, Jun 1898, *Prain s.n.* (L); Goalpara Dist., 26 mi from Dudhnai, 25 Jun 1964, *Rao s.n.* (L); Deaula?, in 1835, *Wallich s.n.* (NY). BIHAR: Without specific locality and date, *Clarke 26783A* (K). MADHYA: Surguja State, Nawadih, among rocks nr. stream, 3600 ft, 10 May 1940, *Mooney 1309* (K). **Madras:** Kodaikanal, edge of lake, 3 Aug 1926, *Bembower 340* (MO); Kodaikanal, Priests Walk, 31 Aug 1929, *Bembower 300* (MICH); Nilgiris Dist., Jun 1883, *Gamble 12051* (K); Nilgiris, 7000 ft, in 1884, *Gamble 15377* (K); Mt. Nilgiri, in 1851, *Hohenacker 969* (L, NY, P); Nilgiri, in 1840, *Perrottet 501* (P); Madura Dist., Palni Hills, in 1914, *Saulière 62* (P). **Punjab:** Kulu, Bijnath?, without date, *Edgeworth 8052* (K). **Sikkim:** Dulkajhar, Terrai, 14 Oct 1884, *Clarke 36946C* (NY). **Uttar:** Rhim Tal, Kumaon, 4400 ft, without date, *Strachey & Winterbottom 7* (GH, K, NY). **West Bengal:** Silligori (Siliguri), 14 Nov 1875, *Kuntze 6468* (NY).

INDONESIA. **Borneo:** Banjarblitong, without date, *Teysmann 8524* (L). **Java:** Without specific locality, date, and collector's name (L). **Moluccas:** Boeroe, 950 m, 22 Feb 1922, *Toxopeus s.n.* (L). **Sumatra:** Si Makkoek, at the high plateau level on the Toba Trail N of the Asahan River, 23 Apr 1923, *Bartlett 7466* (L, MICH, NY, US); Toba, Balige, ca. 1000 m, 14 Mar 1896, *Ouwehand 124* (L); Atjeh, Takengon, 1180 m, 28 Aug 1934, *Van Steenis 5837* (GH, L, NY).

NEPAL. Pokhara, on banks of paddy fields, 9 Jul 1954, *Sykes & Williams 7110* (A, E, US), Argam, nr. Pokhara, 2500 ft, 9 Nov 1954, *Sykes & Williams 7179* (A, E).

SRI LANKA. Without specific locality and date, *Thwaites CP3348* (K).

THAILAND. Loie, Phu Krading, 1300 m, 28 Jun 1948, *Dee 44* (MO); Me Tun, Chengmai, 7 Apr 1922, *Kerr 6236* (L).

VIETNAM. South Vietnam, 28 Feb 1967, *Cuong 159* (P); Annam, Prov. Lang-Biang, 20 Jun 1922, *Poilane 3907* (P).

11. Polygonum breviochreatum Makino, Bot. Mag. (Tokyo) **17:** 117. 1903. *Persicaria breviochreata* (Makino) Ohki, Bot. Mag. (Tokyo) **40:** 53. 1926. *Truellum breviochreatum* (Makino) Sojak, Preslia **46:** 145. 1974. Type. Japan. Kochi: Prov. Tosa, Idzuma, 17 Oct 1892, *Makino 13025* (lectotype, MAK!, designated by Park, Brittonia **38:** 402. 1986; isolectotype, TI!). Syntypes. Japan. Kochi: Prov. Tosa, Sakawa, in 1884, *Makino s.n.* (not seen). Nara: Prov. Yamato, Nara, Oct 1902, *Tsuzi s.n.* (MAK!).

Persicaria ramosa Nakai, Bot. Mag. (Tokyo) **44:** 518. 1930. Type. Japan. "In rupibus secus vias Itsukaichi, prov. Musashi," Sep 1929, *Nakai s.n.* (holotype, TI!).

Erect to scandent *herb,* 0.2–1 m tall; stem light green, slender, relatively weak, branched, glabrous, angled, decumbent at the base, often producing roots at lower nodes, almost smooth to sparsely armed with rather bristlelike recurved prickles along the angles; prickles 1–2.5 mm long. *Leaves* short-petioled, upper ones usually subsessile; blades thin, lanceolate to narrowly elliptic, 2.8–8 cm long, 0.8–1.9 cm wide, acute to acuminate at the apex, at the base weakly cordate to subsagittate with small triangular lobes, sometimes weakly truncate, antrorsely ciliate at the margin; upper surfaces green, glabrous, often bearing short appressed simple multiseriate hairs toward the apex; lower surfaces light green, usually glabrous but often very sparsely pubescent with 5- to 7-rayed sessile stellate hairs; midveins usually antrorsely prickly with bristlelike prickles ca. 0.5 mm long both above and below; petioles short, 0.2–1 cm long, glabrous. *Ocreae* short, 2–5 mm long, scarious, cylindrical, somewhat truncate at the apex, densely hispid with appressed bristles, long-ciliate; bristles 2.5–4 mm long. *Inflorescences* terminal or in the axils of leaves, commonly 2–3 inflorescences arising from an axil, elongate, very slender, often branched, antrorsely hispid, often glandular-pubescent; glandular hairs stalked, ca. 0.5 mm long; inflorescence bracts somewhat reduced but otherwise resembling ocreae; each inflorescence branch terminated by a very loose, interrupted panicle of 2–4 flower fascicles. *Flower fascicles* widely spaced, 2- to 3-flowered; bracts elliptic, 2–2.5 mm long, acuminate at the apex, hispid, long-ciliate; bracteoles lanceolate to elliptic, ca. 3 mm long, glabrous, often bearing a few short simple multiseriate hairs at the apex; pedicels short, 1.5–3 mm long, glabrous, almost included in bracteoles; perianth 5-parted, white with a reddish tinge, 3–3.5 mm long; tepals elliptic, obtuse, not becoming fleshy; stamens seven, rarely six, inserted at the base of perianth in two whorls; inner stamens two, ca. 1 mm long; outer stamens usually five, ca. 0.5 mm long; style one, deeply 3-cleft, ca. 0.5 mm long, included; stigmas three. *Achenes* yellow-brown to brown, ovoid-ellipsoidal, weakly trigonous at the apex, 2–3 mm long, shiny.

Distribution. Known only from Japan (Honshu, Shikoku and Kyushu) and southern Korea (Fig. 24).

Steward (1930) treated *P. breviochreatum* as a taxonomic synonym of *P. strigosum* var. *muricatum* (≡*P. muricatum*). However, it is distinct from *P. muricatum* in having very short densely hispid ocreae, and a loose interrupted inflorescence composed of two to four widely spaced flower fascicles. Other distinguishing features include very slender stems, small, narrowly lan-

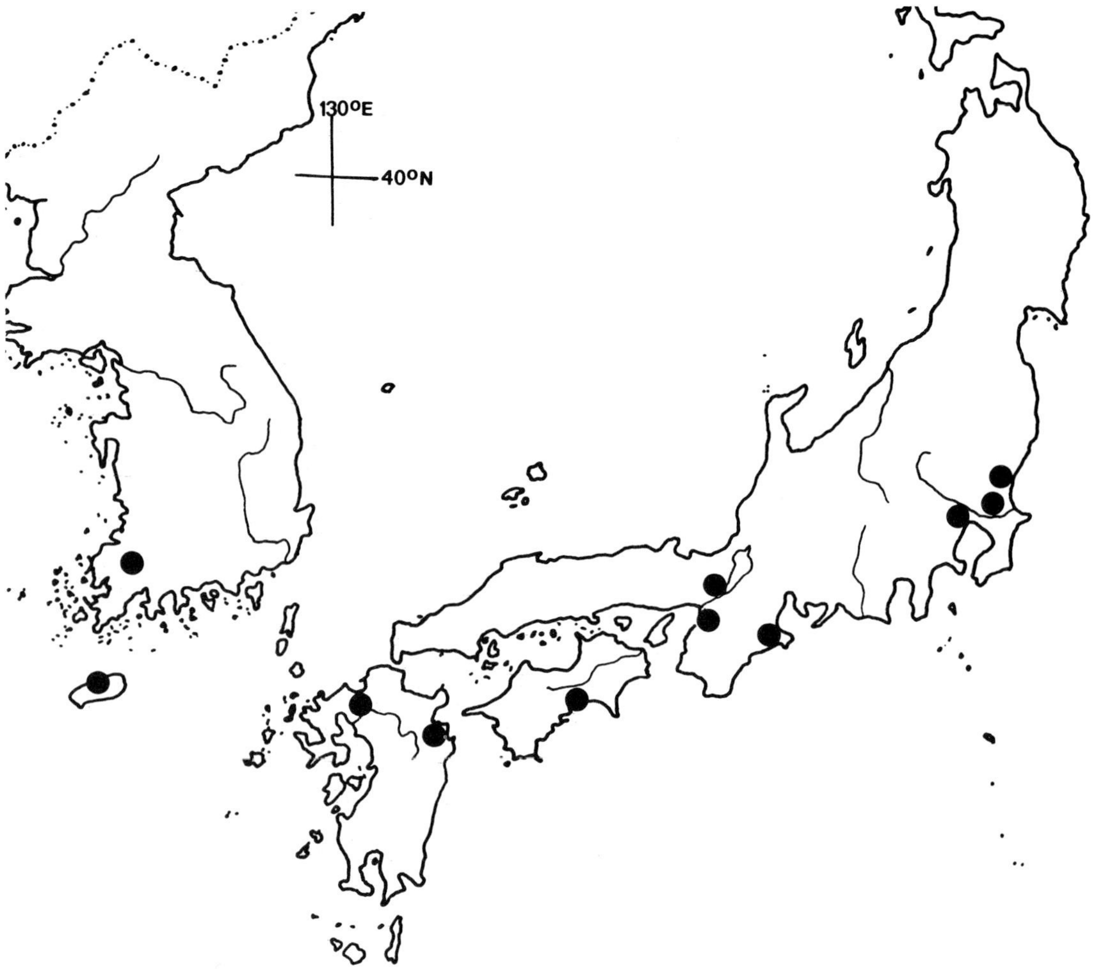

Fig. 24. Distribution of *P. breviochreatum.*

ceolate to elliptic leaves, and ovoid-ellipsoidal achenes.

Additional specimens examined. JAPAN. FUKUOKA: Prov. Chikuzen, Mt. Wakasugi, in wet place along the *Cryptomeria* forest, 500 m, 20 Oct 1952, *Hatusima s.n.* (BH). IBARAGI: Mt. Tsukubasan, Hitachi, 28 Aug 1928, *Hisauchi s.n.* (TUS); Tsukuba-gun, Tsukuba-yama, 1 Oct 1911, *Makino 13005* (MAK). KYOTO: Kyoto City, Sakyo-ku, Kurama-yama, without date, *Makino 13014* (MAK); Furuichi in Tanba, 5 Oct 1957, *Togashi 1635* (BH, LE, NY, TUS, US). MIE: Yuno-yama in Ise, 28 Sep 1950, *Togashi & Matsuoka 177* (BH, E, F, L, MICH, MT, TUS); Yoshino-gun, Ise-gamine, 20 Oct 1895, *Umemura s.n.* (MAK 13010). NARA: Kitakatsuragi-gun, Kashiba-cho, Hata, on the edge of *Cryptomeria* forest, 150–200 m, 6 Oct 1975, *Seto 21S74* (MAK). OITA: Oita-shi, 3 May 1905, *Makino 13019* (MAK). SHIGA: Ohtsu-shi, Mt. Hiei-zan, Anrakuritsuin (300 m)-Yokawachudo (700 m), 15 Sep 1982, *Ohashi et al. 8955* (TUS). TOKYO: Prov. Musashi, Mt. Karyose in Itsuka-ichi, 24 Aug 1929, *Hisauchi s.n.* (TI).

KOREA. CHUNNAM: Kwang-ju, Mt. Mu-deung, 15 Aug 1934, *Smith s.n.* (US). CHEJU: Cheju Isl., Mt. Halla, 21 Aug 1977, *Park 136* (SNU).

12. Polygonum dissitiflorum Hemsley, J. Linn. Soc., Bot. **26:** 338. 1891. *Persicaria dissitiflora* (Hemsley) Gross ex Mori, Enum. pl. Cor. 131. 1922. Type. China. Chekiang: Ningpo Mts., without date, *Faber s.n.* (lectotype, K!, designated by Park, Brittonia **38:** 402. 1986). Syntype. China. Shingking: Without specific locality, Feb 1883?, *Ross s.n.* (K!).

Polygonum fauriei Léveillé & Vaniot *in* Léveillé, Bull. Soc. Bot. France **51:** 423. 1904. *Persicaria fauriei*

(Léveillé & Vaniot) Nakai ex Mori, Enum. pl. Cor. 131. 1922. Type. Korea. "In uliginosis Corea mediae," 6 Sep 1901, *Faurie 570* (holotype, E!).

Polygonum glanduliferum Nakai, J. Coll. Sci. Imp. Univ. Tokyo **23:** 20, fig. 1. 1908. Type. Korea. Kyungbuk: Mt. Cho-ryong, 2 Oct 1902, *Uchiyama s.n.* (not seen; inquiry to TI and other major herbaria in Japan failed to locate the type); see Nakai, J. Coll. Sci. Imp. Univ. Tokyo **31:** 171. 1911 (cited as a syn. of *P. fauriei*).

Ascending to erect *herb,* 0.3–1.5 m tall, somewhat woody at the base; stem branched, angled, often reddish-purple at maturity, sparsely armed with recurved prickles to 1 mm long, often moderately to densely pubescent with many-rayed sessile stellate hairs; rays of stellate hairs ca. 0.1 mm long. *Leaves* long-petioled; blades thin, broadly elliptic to ovate, 4.5–15 cm long, 2–9.5 cm wide, acuminate at the apex, cordate to weakly hastate with two small triangular lobes at the base; middle lobe usually broadly elliptic; margins antrorsely ciliate, upper surfaces green to dark green, sparsely to densely pubescent with stellate hairs identical to those of stem, also bearing some appressed bristlelike simple multiseriate hairs toward the apex; lower surfaces light green, sparsely to moderately stellate-pubescent; stellate hairs sessile, 4- to 8-rayed; rays longer than those of stem and upper surfaces, ca. 0.5 mm long; midveins and major veins on both surfaces usually stellate-pubescent; midveins beneath usually antrorsely prickly with bristlelike prickles to 1 mm long; petioles 1.2–7.4 cm long, stellate-pubescent, often with bristlelike recurved prickles to 2 mm long. *Ocreae* scarious, cylindrical at first but becoming obliquely opened at maturity, 0.5–1.5 cm long, glabrous, ciliate at the tip, lower ones often lacerate. *Inflorescences* terminal or in the axils of upper leaves, elongate, 8.5–30 cm long, branched many times, often geminate, usually moderately to densely pubescent with reddish-purple glandular hairs intermixed with varying densities of simple multiseriate hairs to 1.5 mm long in upper portion, stellate-pubescent and hispid below; stellate hairs similar to those of stem; glandular hairs stalked, 0.5–1 mm long; inflorescence bracts somewhat funnelform, ca. 2 mm long, strongly ciliate, upper ones much reduced; each inflorescence branch terminated by a very loose interrupted panicle of 3–4 flower fascicles. *Flower fascicles* widely spaced, 1- to 2-flowered; bracts somewhat green, funnelform, oblique, 2.5–3 mm long, glabrous, strongly ciliate at the margin, surrounding the inflorescence branch; bracteoles broadly elliptic, 2.5–3 mm long, glabrous; pedicels short, 2–3 mm long, glabrous, almost included in bracteoles; perianth 5-parted, whitish-pink to pink, ca. 3 mm long; tepals broadly elliptic to ovate, not becoming fleshy; stamens 7–8, inserted at the base of perianth in two whorls; inner stamens 1.2–1.5 mm long; outer stamens 1–1.5 mm long; style one, deeply 3-cleft, 1–1.5 mm long, included; stigmas three. *Achenes* spheroidal, weakly trigonous at the apex, dark brown, somewhat dull, 3–3.5 mm long.

Distribution. In moist situations, usually in partial shade; occurring naturally in damp places in forests and along streams, often in roadside ditches; known only from Korea, Manchuria and eastern China; relatively common in Korea (Fig. 25).

Léveillé (1904) distinguished *P. fauriei* from *P. dissitiflorum,* on the basis of its densely glandular-pubescent inflorescences. However, the pubescence of inflorescences in this taxon shows intraspecific and age-related variation; the inflorescences of immature or poorly developed plants are usually lacking glandular hairs and pubescent only with simple multiseriate hairs and stellate hairs. Therefore, *P. fauriei* can hardly be considered as distinct from *P. dissitiflorum.*

Additional specimens examined. CHINA. ANHWEI: Wangshan, in 1973, *Chow 119* (NY); Chiuhwashan, 30 Jul 1934, *Fan & Li 10* (E); Chiuhwashan, 21 Aug 1933, *Sun 1271* (NY). CHEKIANG: Mo Kan Shan, along mountain side, 28 Jul 1926, *Cheo & Wilson 12774* (NY); Tien Mo Shan, 30 Sep 1925, *Ching 9091* (E). KIANGSI: Lushan Mts., 700–800 m, 29 Jul 1933, *Chung & Sun 716* (NY). LIAONING: Fengtien (Mukden), Chienshan, 200 m, 31 Jul 1930, *Kung 860* (NY); Prov. Mukden, "Fluvium Jalu circa Mao-ell-schan," 5 Sep 1897, *Komarov 556* (NY); Prov. Mukden, "in valle Tudagou, supra Mao-ell-schan -Fl. Jalu," 5–17 Sep 1897, *Komarov s.n.* (LE).

KOREA. "In valle Fl. Jalu infra appidulum Chudshu (Korea sept.)," 15 Aug 1897, *Komarov s.n.* (LE). CHUNBUK: Mt. Nae-jang, 6 Oct 1972, *Col.?* (SNU). CHUNGBUK: Mt. Sok-ni, nr. Bub-ju Temple, 26 Aug 1969, *Migo s.n.* (TUS 42324). KANGWON: Mt. Kumgang, 11 Sep 1933, *Chung s.n.* (SNU 1571); Mt. Odae, 25 Jul 1948, *I. Chung 3669* (F, MICH); Mt. Odae, along roadside nr. Wol-jeong Temple, 3 Aug 1981, *Park 661* (BH); Mt. Kum-gang, 28 Jun 1934, *To s.n.* (SNU 1572); Hyangro-bong, 30 Jul 1961, *Yang s.n.* (MAK 44608); Sokumkang, 8 Oct 1971, *Col.?* (SNU); Mt. Chi-ak, 29 Sep 1972, *Col.?* (SNU). KYUNGGI: Kwang-reung, 15 Oct 1947, *Cho 3097* (MICH); Kwang-reung, 15 Oct 1947, *I. Chung 3096* (MICH); Mt. Myung-

Fig. 25. Distribution of *P. dissitiflorum.*

ji, 500 m, 12 Sep 1982, *Kim s.n.* (SNU); Kwang-reung, in wet, shady places along roadside, 22 Jul 1981, *Park 649* (BH); Kwang-reung, wet places nr. stream, 22 Jul 1981, *Park 650* (BH).

13. Polygonum praetermissum Hook. f., Fl. Brit. India **5:** 47. 1886. *Tracaulon praetermissum* (Hook. f.) Greene, Leafl. Bot. Observ. Crit. **1:** 22. 1904. *Persicaria praetermissa* (Hook. f.) Hara, Fl. East. Himalaya 73. 1966. *Truellum praetermissum* (Hook. f.) Sojak, Preslia **46:** 148. 1974. Type. India. Assam: Khasi Hills, 4000 ft, 3 Nov 1872, *Clarke 18212A* (lectotype, K!, designated by Park, Brittonia **38:** 402. 1986). Syntypes. Bangladesh. Silhet, without date, *Wallich 1697-3* (K!). India. Assam: Khasi Hills, Maophlang, 5500 ft, 18 Oct 1872, *Clarke 18584B* (K!); "Khasia Mts.," without date, *Hook. f. & Thomson s.n.* (one sheet believed to be this collection at K). Sri Lanka. Ceylon, 6–7000 ft, *Walker CP2250* (not seen).

Polygonum muricatum var. *auriculatum* Meisner *in* DC., Prodr. **14:** 133. 1856. Type. India. "In itinere Assam," *Griffith s.n.* (not seen; see Steward, Contr. Gray Herb. **5(88):** 86. 1930).

Polygonum birmanicum Gage, Rec. Bot. Survey India **2:** 412. 1903. Types. Burma. Shan Hills, *King's collector 640* (not seen); Fort Stedman, *King's collector 469* (not seen); Thamakhan in southern Shan State,

King's collector s.n. (not seen); see Steward, Contr. Gray Herb. **5(88):** 86. 1930.

Polygonum auriculatum Makino, Bot. Mag. (Tokyo) **17:** 117. 1903; non Meisner, 1826. *Polygonum hastatoauriculatum* Makino ex Nakai, Bot. Mag. (Tokyo) **23:** 420. 1909. *Persicaria hastatoauriculata* (Makino ex Nakai) Gross ex Nakai, Fl. Saishu & Kwan Isl. 41. 1914. *Truellum hastatoauriculatum* (Makino ex Nakai) Sojak, Preslia **46:** 146. 1974. Type. Japan. Kochi: Prov. Tosa, Nakamura-shi, Enomura in Hata-gori, 5 Aug 1889, *Makino 13838* (lectotype, MAK!, designated by Park, Brittonia **38:** 402. 1986). Syntypes. Japan. Aichi: Prov. Migawa, Kameyama in Atsumi-gori, 27 Oct 1893, *Makino 13841 & 13842* (both MAK!); nr. Toyohashi, 29 Oct 1894, *Makino 13844* (MAK!); Hadzu-gori, Noba, 5 Sep 1897, *Nagura s.n.* (not seen). Chiba: Nagara-gori, Nakahara-mura, Sep 1880, *Col.?* (not seen). Ehime: Onsen-gori, Shoyenzi-mura, Jul 1894, *Okudaira s.n.* (not seen). Gifu: Prov. Mino, Iwato-mura, without date, *Yatabe & Matsumura s.n.* (not seen). Kochi: Prov. Tosa, Erasawa, 15 Aug 1890, *Doi s.n.* (not seen); Iwado, 28 Aug 1890, *Doi s.n.* (not seen); Kochi-shi, Asakura, 15 Oct 1892, *Makino 13840 & 13843* (both MAK!); Tosa-shi, Takaoka-gun, Ryumura, 16 Oct 1892, *Yano s.n.* (MAK!). Miyazaki: Prov. Hyuga, Takanabe, 27 Jul 1822, *Yatabe & Matsumura s.n* (not seen). Ryukyu: Kume-shima, Uizatomagiri, in 1898, *Kuroiwa s.n.* (MAK!). Yamaguchi: Prov. Suo, Ouchi-mura, 10 Jul 1892, *Nakai s.n.* (not seen). Taiwan. Taipei, 10 Nov 1896, *Makino s.n.* (not seen).

Polygonum thunbergii var. *hastatotrilobum* subvar. *eciliolatum* Léveillé, Bull. Soc. Bot. France **51:** 423. 1904. Type. Not cited (see Lauener, Notes Roy. Bot. Gard. Edinburgh **40:** 200. 1982).

Scandent *herb* arising from long decumbent radicant base, 0.3–1 m tall; stem often branched, angled, sparsely to moderately armed with recurved prickles along the angles; prickles bristlelike, 1–1.5 mm long. *Leaves* short-petioled, upper ones often subsessile; blades linear lanceolate to narrowly elliptic, 3.5–8.5 cm long, 0.5–1.7 cm wide, often elliptic to ovate in lower ones, glabrous on both surfaces, gradually attenuate into obtuse to subacute tip, weakly hastate to subsagittate at the base but upper ones often shallowly cordate; basal lobes oblong to triangular, 0.3–1 cm long, obtuse to acute, often deflexed; margins finely ciliate; midveins beneath sparsely prickly; petioles short, 0.5–1 cm long, with bristlelike prickles to 1.5 mm long. *Ocreae* scarious, cylindrical, 1–1.7 cm long, oblique at the apex, usually glabrous but often with varying densities of short appressed bristlelike hairs along the veins, often short-ciliate, bearing a ring of recurved prickles at the base, lower ones usually lacerate. *Inflorescences* terminal or in the axils of upper leaves, elongate, slender, 5–8 cm long, branched 1–3 times, often geminate, glabrous to sparsely pubescent with glandular hairs in upper portion; glandular hairs stalked, ca. 0.5 mm long; inflorescence bracts resembling ocreae but much reduced, somewhat funnelform, 1.5–2.5 mm long; each inflorescence branch terminated by a loose interrupted panicle of 3–7 flower fascicles. *Flower fascicles* widely spaced, 2- to 4-flowered; bracts funnelform, oblique, 2–2.5 mm long, glabrous, often ciliate, surrounding the inflorescence branch; bracteoles elliptic, ca. 3 mm long, glabrous; pedicels short, 3–4 mm long, glabrous, slightly exserted at maturity; perianth 4-parted, reddish, 3–4 mm long; tepals broadly elliptic, obtuse, not becoming fleshy; stamens 4–5, inserted at the base of perianth in one whorl, 1–1.5 mm long; style one, 3-cleft up to middle, ca. 0.5 mm long, included; stigmas three. *Achenes* spheroidal, weakly trigonous at the apex, 2–3 mm long, dark brown to black, dull, rugose.

Distribution. From northern India through China and southern Korea to Japan, and southward to southeastern Australia through Ryukyu Islands and Philippines (Fig. 26).

Japanese populations have been described as a distinct species, *P. auriculatum,* by Makino (1903). Nakai (1909) later changed the name to *P. hastatoauriculatum,* recognizing that Meisner (1826) had validly published the name *P. auriculatum* for a different species. After examining the type specimens and other specimens from Japan, it is evident that *P. hastatoauriculatum* does not differ from *P. praetermissum* (Park, 1986b).

Collections from southeastern Australia differ from the rest of the species in having long-petioled narrowly triangular leaves with small, acute to acuminate, triangular basal lobes. Petioles are distinctive and much longer, as compared to those of other populations, and range from $\frac{1}{10}$ to $\frac{3}{10}$ of the length of blade. In addition, pedicels are pubescent with stalked glandular hairs and they are somewhat exserted at maturity. These populations occupy the southern extreme of the distribution range and may represent an initial geographical speciation. These plants are frequently confused with *P. strigosum,* but they can be readily distinguished by their four-parted perianths and rugose achenes.

Additional specimens examined. AUSTRALIA. NEW SOUTH WALES: Parramatta, Dec 1807, *Caley s.n.* (K); Parramatta, in 1810, *Caley s.n.* (US); Liston, head of

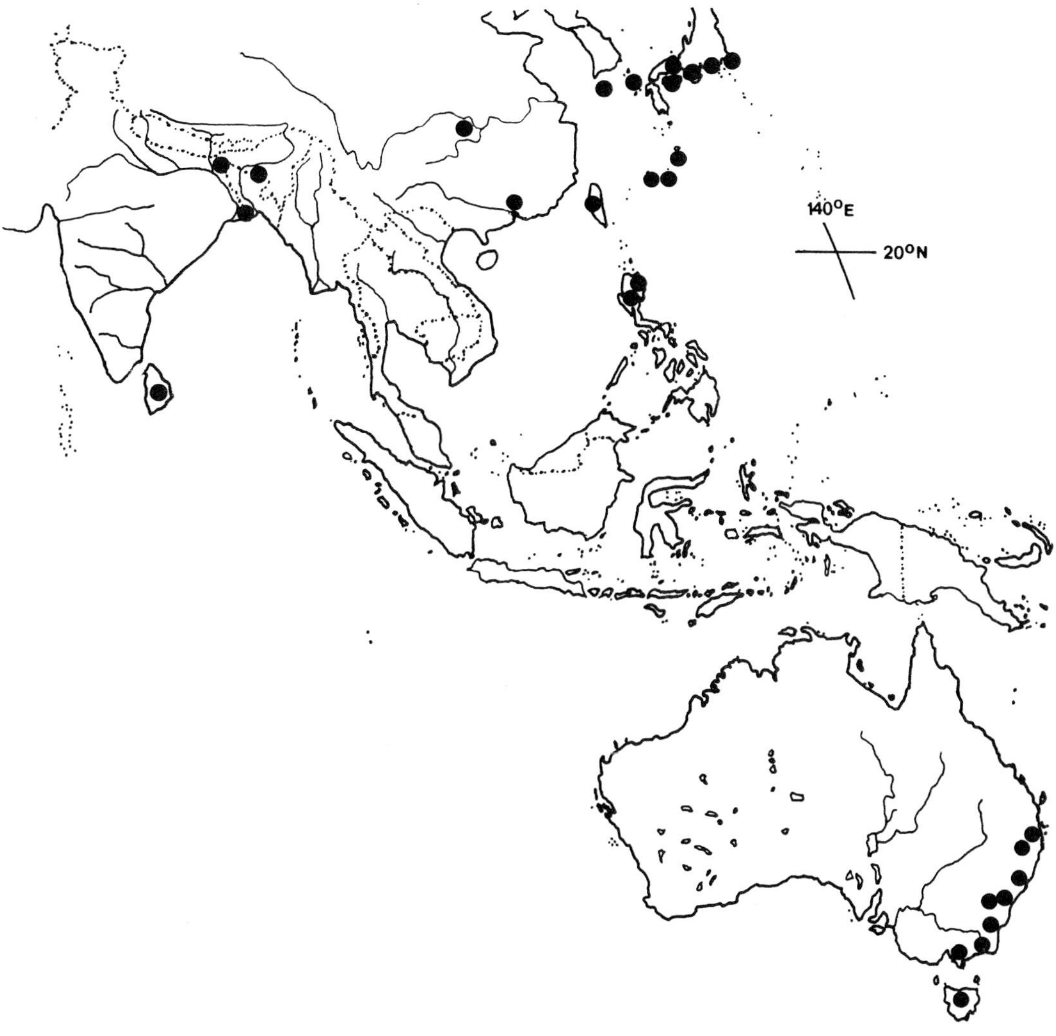

Fig. 26. Distribution of *P. praetermissum.*

Wylie Gorge, 17 Nov 1943, *Clemens s.n.* (GH); 3.4 km by rd. N of Bodalla P.O., 17 Dec 1975, *Coveny 7325* (K); 7.2 km by rd. NNE of Moruya Police Station, 17 Dec 1975, *Coveny 7335* (K); Lisarow, 15 m, 10 Jun 1975, *Coveny & Powell 6463* (K); Wingecarribee Swamp, behind Burrawang Railway Station, 7 Mar 1976, *Wilson 1454* (K); ca. 1.5 km S of Wyong Creek and 0.5 km W of Pacific Hwy., S of Wyong, 12 Dec 1978, *Wilson 1988* (K); Cox River on Lowther-Little Hartley Rd., 23 Jan 1979, *Wilson 2023* (K); ⅓ km along Diamond Head Rd. from Mooreland, 18 Dec 1981, *Wilson 3917* (K); 1 km along Toonumber Forest, way from Kyogle-Toonumber Dam Rd., 28 Dec 1981, *Wilson 4060* (K). **Queensland:** Sunny bank nr. Brisbane, 130 ft, 23 Apr 1930, *Hubbard 2353* (K). **Tasmania:** Launceston, 26 Mar 1842, *Gunn s.n.* (K). **Victoria:** Lysterfield-Emerald, about 4¼ mi ESE of Lysterfield, at Muddy Creek where crossed by the Lysterfield-Emerald Rd., 2 Dec 1963, *Aston 924* (K); Gippsland, appr. 16 km N of Maffra, Avon River at Valencia Creek Township, 10 May 1979, *Aston 2023* (K); South Gippsland, Tarra River, about 3 mi SE of Yarram, 28 Apr 1964, *Jackson 149* (A).

CHINA. **Hupeh:** Without specific locality, 1885–1888, *Henry 1669* (US); Ichang, without date, *Henry s.n.* (K). **Kwangtung:** Canton vic., 22 Aug 1918, *Levine 2155* (MO, US); Shin Cheng Sha Tan, 12 Apr 1914, *Tutcher s.n.* (K).

INDIA. **Assam:** Mt. Khasia, without date, *Hook f. & Thomson s.n.* (L, US); Cherrapunjee, 4000 ft, 30 Apr 1952, *Koelz 29563* (L); Mawphlang (Maophlang), 6000 ft, 11 Sep 1952, *Koelz 31258* (L); Prov. Khasia, Tserapindz till Mairang, 2800–4500 ft, 1–30 Oct 1855, *Schlagintweit 386* (NY). **Bihar:** N. Purnea, 29 Oct 1918, *Haines 4599* (K). **West Bengal:** Ganges Plain, without date, *Hook f. & Thomson s.n.* (L, NY).

JAPAN. CHIBA: Prov. Kazusa, Mobara, in 1924, *Makino 13879* (DAO, E, LE). FUKUOKA: Fukuoka City, Shingu, 24 Sep 1933, *Hatusima s.n.* (US). GIFU: Ohgaki, 30 m, 25 Jun 1971, *Kanai & Morita 283* (TUS). HYOGO: Arima, Nanakusa-yama, Nov 1962, *Col.?* (MAK 13863). KAGAWA: Takamatsu-shi, Hirai, 15 Sep 1912, *Matsubara s.n.* (MAK 13869). KAGOSHIMA: Satsuma, Lake Satsumako, nr. Izaku, sea level, 4 Sep 1956, *Koyama 7055* (NY, TI); Prov. Satsuma, 4 Nov 1923, *Masamune s.n.* (NY). KOCHI: Without specific locality, 16 Oct 1892, *Yano s.n.* (MAK 13837). NAGASAKI: Isl. Fukue-jima, at the NW foot of Mt. Ooto-zan, Yoshida-cho, Fukue City, 23 Aug 1982, *Fukuoka & Kurosaki 3701* (TI); Isl. Iki, in 1950, *Shinagawa s.n.* (TI). OKAYAMA: Prov. Bizen, 7 Oct 1919, *Masamune s.n.* (NY); Okayama City, Niwase, bank of the Ashimori-gawa River, 7 Jun 1982, *Tsuchiya 2276* (TI); Kurashiki-shi, Hukuda, 27 Oct 1893, *Uno s.n.* (MAK 13841); Okayama-shi, 15 Oct 1913, *Uno 4755* (NY). OSAKA: Settsu-shi, Shinzaike, Torigai, 6 Oct 1973, *Seto & Okamoto 7* (K, MSC). RYUKYU: Okinawa Isl.: Kushi, 12 Oct 1955, *Amano 7572* (L); Masatomura, Nakagami, in the pond, 7 May 1955, *Hatusima 17622* (US); Goyekku-mura, Nakagami, in the pond in the settlement of Uchikina, 18 May 1955, *Hatusima 17883* (L), in the wet grass field nr. the City of Nago, Kunigami, 19 May 1955, *Hatusima 18055* (US); Onnamura, Nakagami, in the brook along the paddy field, 8 May 1956, *Hatusima 17713* (TI); Kushison, 8 Dec 1938, *Kanashiro s.n.* (US); Koza, Goeku, 19 Oct 1958, *Koyama 7303* (NY); Nakagami, Uchikina, Goeku-son, 27 Jun 1951, *Walker 5605* (MICH); Nakagami, Uchikina, Goeku-son, 27 Jun 1951, *Walker & Amano 5905* (L, US); Okuma, Kunichamma-giri, in 1904, *Col.?* (NY). Tokunoshima Isl.: in abandoned paddy field at San, 50 m, 18 Aug 1955, *Hatusima 19165* (L). SHIGA: Sakata-gun, Ibuki-san, 3 Nov 1893, *Makino 13858* (MAK). WAKAYAMA: Tanabe-shi, Tanabe, in 1924, *Makino 13854* (MAK).

KOREA. CHEJU: Cheju Isl., Aug 1907, *Faurie 2041* (E); Cheju Isl., Mt. Halla, Aug 1976, *Park 86* (BH); Cheju Isl., Aug 1911, *Taquet 5908* (E); Cheju Isl., in 1912, *Taquet 6204* (TI).

PHILIPPINES. LUZON. Benguet: Palou, without date, *Loher 4598* (K); without specific locality, May 1911, *Merrill 7749* (L); without specific locality, May 1914, *Merrill 1713* (L, NY). Lepanto: Mt. Data, Nov 1905, *Merrill 4521* (K, L, NY).

SRI LANKA. Ceylon, in 1847, *Gardner 724* (K); Ceylon, 6000 ft, in 1854, *Thwaites 2250* (P); Ceylon, without date, *Walker s.n.* (K); Ceylon, without date, *Walker 213* (K).

TAIWAN. Without specific locality, in 1900, *Makino 13877* (MAK). ILAN: Mt. Taiping, 23 Jul 1970, *Kao s.n.* (TAI). PINGTUNG: Pingtung, 13 Sep 1978, *Ou & Kao 9054* (TAI). TAIPEI: Taipei, Jun 1903, *Léveillé s.n.* (E).

14. Polygonum subsagittatum (De Wildeman) Park, Brittonia **38:** 218. 1986. *Polygonum pedunculare* var. *subsagittatum* De Wildeman, Contr. Fl. Katanga, Suppl. 3. 107. 1930. Type. Congo. Fendula (Nyakolonge), in 1928, *Scaetta 641* (holotype, BR; photo of holotype, K!).

Ascending to scandent *herb,* 1–1.5 m tall; stem branched, angled, moderately to densely armed with recurved prickles, also bearing varying densities of minute 2- to 4-rayed tufted hairs, often with a few glandular hairs in upper portion; prickles bristlelike, 1.5–2 mm long; glandular hairs stalked, brown, ca. 0.5 mm long. *Leaves* petioled, upper ones often subsessile; blades linear lanceolate to lanceolate, 6.5–10 cm long, 0.8–1.7 cm wide, acuminate at the apex, weakly hastate to sagittate with small obtuse or acute triangular lobes at the base but upper ones often shallowly cordate to truncate; basal lobes often obliquely deflexed; margins ciliate; upper surfaces sparsely to densely pubescent with appressed bristlelike simple multiseriate hairs ca. 2 mm long, often intermixed with 2- to 4-rayed tufted hairs similar to those of stem; rays ascending, ca. 0.1 mm long; lower surfaces almost glabrous to densely pubescent with 2- to 4-rayed tufted hairs; rays slender, much longer than those of stem and upper surfaces, ca. 0.3–0.5 mm long; midveins and major veins beneath bearing bristlelike prickles, 1–1.5 mm long; petioles relatively short, 0.5–1.5 cm long, retrorsely prickly, often with some tufted hairs identical to those of stem. *Ocreae* scarious, cylindrical, 1.5–2.3 cm long, obliquely opened, short-ciliate at the tip, sparsely to moderately hispid with bristles to 2 mm long along the veins, bearing a ring of recurved prickles at the base; lower ocreae often glabrous and lacerate. *Inflorescences* terminal or in the axils of upper leaves, elongate, 10–16 cm long, branched 1–3 times, often geminate, sparsely to densely pubescent with glandular hairs identical to those of stem; inflorescence bracts resembling ocreae but much reduced, 3–5 mm long; each inflorescence branch terminated by a very loose interrupted panicle of 4–7 flower fascicles. *Flower fascicles* widely spaced, 2- to 3-flowered; bracts somewhat funnelform, 2–2.5 mm long, hispid, ciliate at the margin, surrounding the inflorescence branch; bracteoles broadly elliptic to ovate, 2–2.5 mm long, glabrous, ciliate at the apex; pedicels 2–2.5 mm long, glandular-pubescent, often slightly exserted at maturity; perianth 5-parted, white with a pink tinge, 3–4 mm long; tepals elliptic to broadly elliptic, not becoming fleshy; stamens five, inserted at the base of peri-

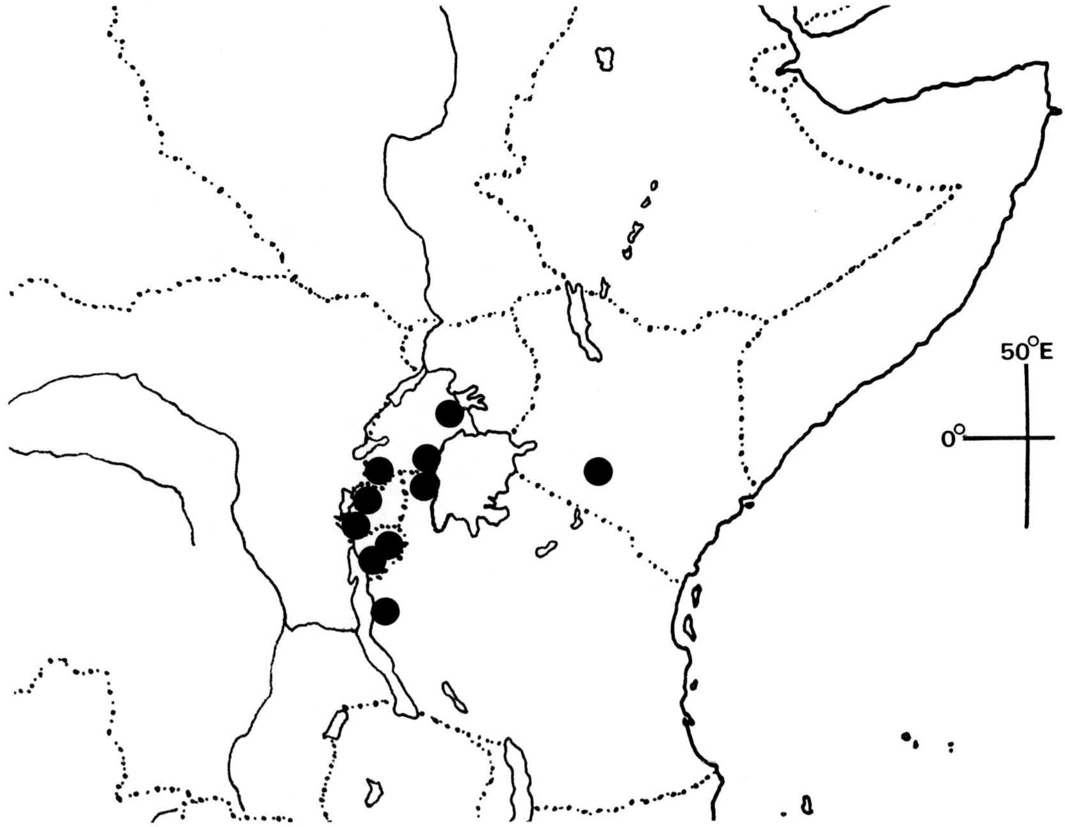

Fig. 27. Distribution of *P. subsagittatum.*

anth in one whorl, 0.8–1 mm long; style one, 2-cleft, ca. 0.5 mm long, included; stigmas two. *Achenes* biconvex, ovate in outline, 2–2.5 mm long.

Distribution. Known only from southeastern Africa; from Uganda to Tanzania along Lake Victoria and Lake Tanzania (Fig. 27).

Polygonum subsagittatum has been recognized as a variety of *P. pedunculare* (=*P. dichotomum*), but it is completely distinct from that species in having loose, interrupted inflorescences (Park, 1986a). In addition, the bracts that subtend flower fascicles completely surround the inflorescence branch, and stems and both surfaces of the leaves are pubescent with two- to four-rayed tufted hairs.

Additional specimens examined. BURUNDI. RUYIGI: Kigamba, 1550 m, 2 Oct 1974, *Auquier 4264* (MO); Messe-Ruyigi, 8 Sep 1951, *Michel & Reed 222* (MO); Messe-Ruyigi, Sep 1951, *Michel & Reed 242* (MO); route Gitega–Kinyinya km 70 marais, 1250 m, 27 Sep 1974, *Reekmans 3714* (MO); Gaterama, 1700 m, 24 Feb 1977, *Reekmans 5697* (K, MO); Gitwenge colline, Mukagoma, 2 Jul 1979, *Reekmans 7632* (K).

KENYA. Mt. Aberdare, Niihi River, in 1925, *Fries 1948* (K).

RWANDA. KALCHE: Zwiere Nyawarongo, 1500 m, 12 Feb 1959, *Leonard 2997* (K, L). SHANGUGU: Rugege, 1920 m, 25 Jul 1974, *Auquier 3406* (MO); Kamiranjovu, 2000 m, 18 Mar 1956, *Christianensen 1411* (K); route Bukavu–Astrida, 1900 m, 23 Jul 1959, *Leonard 5090* (K).

TANZANIA. BUKOBA: Gera Camp, 31 Aug 1934, *Gillman & Marshall 141* (K). KIGOMA: Kasangazi, Mahali Mts., 24 Jul 1958, *Jefford & Juniper 212* (MO).

UGANDA. Banks of Sezibwa River, 4000 ft, 2 Aug 1915, *Dümmer 1084* (K); between Kinanira and Kisolo, 3 Apr 1927, *Linder 2549* (GH); Kigezi, S end of Lake Mutande, 6200 ft, 10 Jan 1933, *Rogers 320* (K); Sango Bay, Katera, 24 Jul 1935, *Thomas 1312* (K).

15. Polygonum bungeanum Turcz., Bull. Soc. Imp. Naturalistes Moscou 13: 77. 1840.

Polygonum pensylvanicum sensu Bunge, Enum. Pl. China Bor. 57. 1833; non L., 1753. *Persicaria bun-*

geana (Turcz.) Nakai ex Mori, Enum. pl. Cor. 131. 1922. Type. China. "In humidis prope Kantai," without date, *Bunge s.n.* (holotype, P!).

Polygonum chanetii Léveillé, Bull. Soc. Bot. France **54:** 370. 1907. Type. China. HOPEH: Tchao-Tchao, 10 Aug 1905, *Chanet 86* (holotype, E!).

Erect to ascending *herb,* 30–80 cm tall; stem branched, angled, sparsely armed with short stout prickles to 1.5 mm long, often bearing a few glandular hairs in upper portion; glandular hairs stalked, ca. 0.5 mm long. *Leaves* petioled; blades lanceolate to narrowly elliptic, 5–12.5 cm long, 1.5–3.5 cm wide, subacute to subacuminate, rarely obtuse at the apex, acute at the base, antrorsely ciliate at the margin; upper surfaces green, almost glabrous, usually antrorsely prickly with fine prickles along the midvein; lower surfaces light green, almost glabrous to moderately pubescent with short simple multiseriate hairs; midveins and major veins beneath finely antrorsely prickly; petioles 0.5–1.5 cm long, finely antrorsely prickly. *Ocreae* scarious, cylindrical, 0.8–1.4 cm long, truncate, long-ciliate, bearing somewhat appressed bristles to 2 mm long along the veins. *Inflorescences* terminal or in the axils of upper leaves, elongate, 5.5–11.5 cm long, often branched 1–2 times, sparsely to densely pubescent with glandular hairs in lower portion; glandular hairs identical to those of stem; inflorescence bracts resembling ocreae but much reduced, somewhat funnelform, 2–6 mm long; each inflorescence branch terminated by a loose interrupted spikelike cluster of 8–10 flower fascicles. *Flower fascicles* relatively widely spaced, 2- to 4-flowered; bracts funnelform, oblique, 2–3 mm long, eciliate, glabrous but often bearing a few glandular hairs, completely surrounding the inflorescence branch; bracteoles somewhat funnelform, glabrous, 2–2.5 mm long; pedicels short, 3–3.5 mm long, glabrous, almost included in bracteoles; perianth 5-parted, pale green, slightly red-tinged, ca. 3 mm long; tepals elliptic to broadly elliptic, obtuse, not becoming fleshy; stamens eight, inserted at the base of perianth in two whorls; inner stamens ca. 1.5 mm long; outer stamens ca. 1 mm long; style one, 2-cleft up to middle, 1–1.5 mm long, included; stigmas two. *Achenes* biconvex, obovate in outline, black, dull, rugose, acute at the apex, ca. 3 mm long.

Distribution. Relatively common in northern China, Manchuria, northern Korea, and Japan (Fig. 28); recently introduced and becoming a successful weed in the southern Minnesota counties of Blue Earth, Brown, Faribault, Martin, Waseca, and Yellow Medicine, and in the northern Iowa county of Winnebago (Andersen et al., 1985).

Polygonum bungeanum closely resembles the species of section *Persicaria.* As in taxa of section *Persicaria,* its flower fascicles are arranged in relatively long spikelike clusters, and leaves are lanceolate to narrowly elliptic with acute to attenuate bases. In addition, the bracts that subtend flower fascicles are funnelform and completely surround the inflorescence branch. However, the presence of recurved prickles on the stem clearly aligns it with section *Echinocaulon.*

Additional specimens examined. CHINA. "Chine septentrionale," without date, *Tsche 83* (P). HOPEH: Sulu Hsien, Hsin Chi, 15 Aug 1948, *Beach 49* (MICH, US); Chihli Province, E of Tientsin, in 1913, *Clemens 8063* (E); Peitaiho, 30 Jun 1933, *Hao 1605* (K); Jehol, 8 Aug 1933, *Nakai et al. 538* (TI); without specific locality, 10 Jun 1958, *Col.?* (PE). KIANGSU: Kiangsu, 27 Nov 1958, *Chow s.n.* (NSM 318823). LIAONING: Liaotung Peninsula, nr. Sta. Vanfangoo, 20 Jul 1902, *Litvinov 1631* (NY); Mukden, 9 Aug 1902, *Litvinov s.n.* (MAK, MICH, US); Mukden to Liaoyang, Feb 1883, *Ross 347* (K). PEKING: Peking City, 18 Jun 1976, *Lee 7602* (PE). SHANTUNG: Without specific locality and date, *Izumizaki s.n.* (MAK 13028); without specific locality, 6 Aug 1920, *Yabe s.n.* (NSM 96506).

JAPAN. AICHI: Nagoya-shi, Jun 1922, *Makino 13027* (MAK). KANAGAWA: Sagami, 12 Jun 1929, *Hisauchi s.n.* (TUS 16682). NAGANO: Matsumoto-shi, Utsukushigahara, 7 Aug 1929, *Makino 13026* (MAK). SAITAMA: Urawa-shi, Aug 1906, *Takayanagi s.n.* (MAK 13029).

U.S.A. MINNESOTA. Blue Earth Co.: NW ¼ NE ¼ Sec. 4 T105N R26W, approx. 0.75 mi W of Mapleton, 9 Aug 1984, *Zaremba s.n.* (BH).

16. Polygonum rubricaule Cham., Linnaea **8:** 130. 1833. *Tracaulon rubricaule* (Cham.) Greene, Leafl. Bot. Observ. Crit. **1:** 22. 1904. *Truellum rubricaule* (Cham.) Sojak, Preslia **46:** 148. 1974. Type. Brazil. *Sellow* (not seen; see Park, Brittonia **38:** 403. 1986).

Ascending to erect *herb,* 0.5–1.5 m tall; stem branched, angled, glabrous, almost unarmed to sparsely armed with recurved prickles, reddish-purple at maturity; prickles short, stout, ca. 1 mm long. *Leaves* short-petioled; blades thin, lanceolate to narrowly elliptic, 3.5–7 cm long, 0.5–1.3 cm wide, acute at the apex, weakly cordate to sagittate with small round to triangular lobes at the base, antrorsely ciliate at the margin, gla-

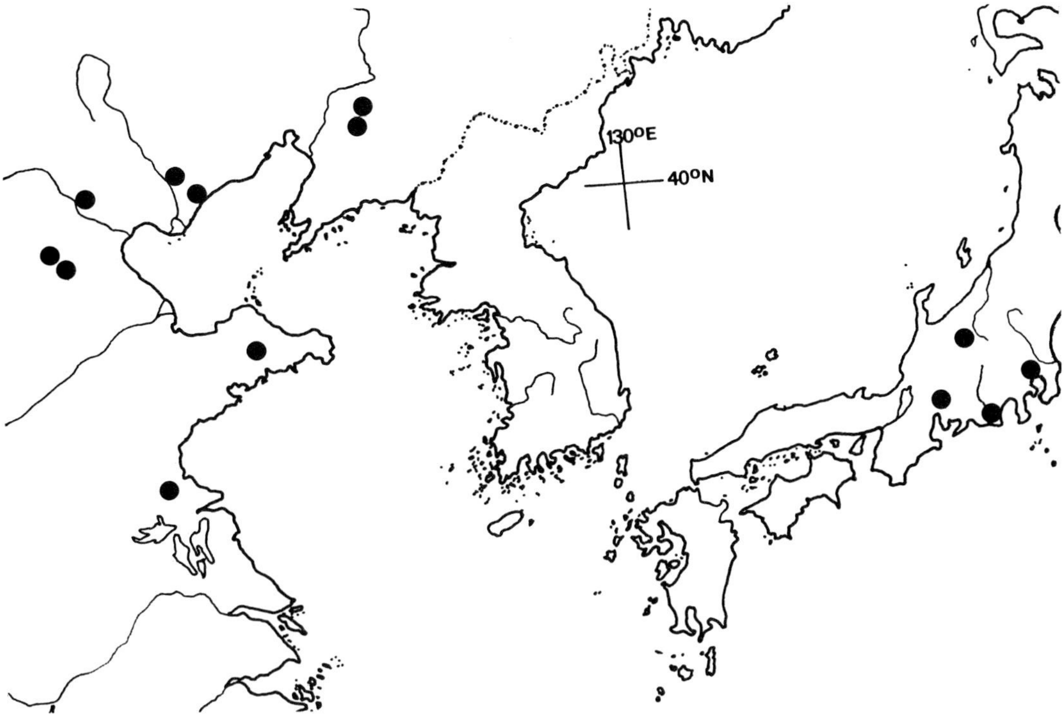

Fig. 28. Distribution of *P. bungeanum.*

brous on both surfaces; basal lobes often obliquely deflexed; midveins beneath armed with recurved prickles to 1 mm long; petioles very short, 2–4 mm long, glabrous. *Ocreae* scarious, 0.8–1.8 cm long, oblique, acute at the apex, glabrous, eciliate, lower ones often lacerate. *Inflorescences* terminal or in the axils of upper leaves, elongate, 5–11 cm long, branched 1–3 times, often geminate, glabrous; inflorescence bracts resembling ocreae, but somewhat reduced and funnelform, often lacerate, 2–3 mm long; each inflorescence branch terminated by a loose, interrupted panicle of 2–4 flower fascicles. *Flower fascicles* widely spaced, 2- to 4-flowered; bracts somewhat funnelform, oblique, acute at the apex, glabrous, 1.5–2 mm long, surrounding the inflorescence branch; bracteoles lanceolate to narrowly elliptic, glabrous, 2–3 mm long; pedicels short, 2–3 mm long, glabrous, almost included in bracteoles; perianth 5-parted, 2–3 mm long; tepals broadly elliptic, not becoming fleshy; stamens eight, inserted at the base of perianth in two whorls; inner stamens three, ca. 1.5 mm long; outer stamens five, ca. 1 mm long; style one, 3-cleft, ca. 1 mm long, included; stigmas three. *Achenes* black, weakly trigonous, ovoid in outline, acute at the apex, smooth, shiny, ca. 3 mm long.

Distribution. In southeastern Brazil, from Minas Gerais to Paraná (Fig. 29); also known from Africa (Tanzania, Mufindi).

Chamisso (1833), in his original description, pointed out that *P. rubricaule* has two-cleft styles. However, Meisner examined the type specimen at B for his Polygonaceae in *Flora Brasiliensis* (1855) and indicated that it has three-cleft styles. All specimens of *P. rubricaule* examined for this study also have three-cleft styles.

Additional specimens examined. BRAZIL. Without specific locality and date, *Glaziou III-1027 & 1028* (P). MINAS GERAIS: Caldas?, 4 Mar 1868, *Henschen 1020* (US). PARANÁ: Fernandes Pinheiro, 26 Mar 1904, *Danser s.n.* (L); Mun. Ponta Grossa, Parque Vila Velha, Rio Guavirova, 11 Mar 1962, *Hatschbach 8983* (L, US); Mun. Curitiba, Barigui, 2 Jan 1971, *Hatschbach 26208* (NY, UEC); Mun. Pirai do Sul, Tijuco Preto, 17 Nov 1970, *Hatschbach & Guimaraes 25395* (K). SÃO PAULO: Without specific locality, in 1883, *Gaudichaud s.n.* (P); Rincao, 25 Jan 1928, *Toledo s.n.* (SP); Ribeirao, Nov 1910, *Col.?* (SP).

TANZANIA. Mufindi, Lulangi, 6000–6500 ft, 19 Dec 1971, *Paget-Wilkes 1013* (MO).

Fig. 29. Distribution of *P. rubricaule* in South America.

17. Polygonum sagittatum L., Sp. pl. 363. 1753. *Helxine sagittatum* (L.) Raf., Fl. tellur. **3:** 10. 1837. *Tracaulon sagittatum* (L.) Small, Fl. S.E. U.S. 381. 1903. *Persicaria sagittata* (L.) Gross ex Nakai, Fl. Saishu & Kwan Isl. 41. 1914. *Truellum sagittatum* (L.) Sojak, Preslia **46:** 149. 1974. Type. Clifford specimen which vouchered tab. 12 in Linnaeus, Hortus Cliffortianus, 1737 (lectotype, BM, designated by Reveal, J. Linn. Soc., Bot. **92:** 168. 1986).

Polygonum sagittatum var. *boreale* Meisner, Monogr. Polyg. 65. 1826. Type. Not cited [see Meisner *in* DC., Prodr. **14:** 132. 1856 (cited as a syn. of *P. sagittatum*)].

Polygonum sagittatum var. *americanum* Meisner *in* DC., Prodr. **14:** 132. 1856. *Persicaria americana* (Meisner) Ohki, Bot. Mag. (Tokyo) **40:** 55. 1926. Type. U.S.A. New York: Schoharie Co., Schoharie Creek, in 1827, *Janin s.n.* [lectotype, G-DC (microfiche at NY!), designated by Park, Brittonia **38:** 403. 1986]. Syntypes. U.S.A. Without specific locality, in 1831, *Rafinesque s.n.* [G-DC (microfiche at NY!)]. Ohio: Without specific locality and date, *Frank 1583* [G-DC (microfiche at NY!)]. "Carol. med.," without date, *Fraser 192* [G-DC (microfiche at NY!)].

Polygonum sagittatum var. *sibiricum* Meisner *in* DC., Prodr. **14:** 132. 1856. *Tracaulon sibiricum* (Meisner) Greene, Leafl. Bot. Observ. Crit. **1:** 22. 1904. *Truellum sibiricum* (Meisner) Sojak, Preslia **46:** 148. 1974. Type. U.S.S.R. Irkutsk, in 1843, *Schschukin s.n.* [lectotype, G-DC (microfiche at NY!), designated by Park, Brittonia **38:** 403. 1986].

Polygonum sieboldii Meisner *in* DC., Prodr. **14:** 133. 1856. *Polygonum sagittatum* var. *sieboldii* (Meisner) Maxim. ex Komarov, Trudy Imp. S.-Peterburgsk. Bot. Sada **22:** 132. 1903. *Tracaulon sieboldii* (Meisner) Greene, Leafl. Bot. Observ. Crit. **1:** 22. 1904. *Persicaria sagittata* var. *sieboldii* (Meisner) Nakai, Fl. Saishu & Kwan Isl. 41. 1914. *Persicaria sieboldii* (Meisner) Ohki, Bot. Mag. (Tokyo) **40:** 54. 1926. *Polygonum anguillanum* Koidzumi, Acta Phytotax. Geobot. **8:** 51. 1939 (see Sojak, Preslia **46:** 149. 1974). *Persicaria anguillana* (Koidzumi) Honda, Bot. Mag. (Tokyo) **53:** 383. 1939. *Truellum sieboldii* (Meisner) Sojak, Preslia **46:** 149. 1974. Type. Japan. Without specific locality, in 1847, *Zollinger 207* [holotype, G-DC (microfiche at NY!); isotype, A!].

Polygonum sagittatum var. *hallaisanense* Léveillé, Repert. Spec. Nov. Regni Veg. **8:** 171. 1910. Type. Korea. Cheju: "Quelpaert, in humidis Hallaisan," 1700 m, 7 Sep 1908, *Taquet 1313* (lectotype, E!, designated by Park, Brittonia **38:** 403. 1986). Syntype. Korea. Cheju: "Quelpaert, in herbidis Hallaisan," 1700 m, 13 Aug 1908, *Taquet 1296* (K!).

Polygonum sagittatum var. *ovalifolium* Farwell, Pap. Michigan Acad. Sci. **3:** 95. 1924. Type. U.S.A. Michigan: Oakland Co., Oxford, 23 Aug 1922, *Farwell 6369* (holotype, MICH!).

Polygonum belophyllum Litvinov, Spisok. Rast. Gerb. Fl. SSSR Bot. Inst. Vsesojuzn. Akad. Nauk **9:** 25. 1932. Type. U.S.S.R. Transbaicalia: Dist. Czita, "in prato paludoso pr. acidulum Makowejewsky," 27 Jul 1909, *Stukov s.n.* (isotype?, K!).

Polygonum sagittatum var. *gracilentum* Fern., Rhodora **44:** 393. 1942. Type. U.S.A. Virginia: Essex Co., brackish marsh along Piscataway Creek, NW of Dunnsville, 1 Aug 1941, *Fernald & Long 13331* (holotype, GH!; isotype, MO!).

Scandent annual *herb,* 0.3–2 m tall; stem light green to green, often becoming reddish-purple at maturity, simple to extensively branched, glabrous, angled, armed with recurved prickles along the angles, often decumbent at the base, usually producing fibrous roots at lower nodes; prickles stout, broad-based, 1–1.5 mm long. *Leaves* petioled, upper ones often short-petioled or subsessile; blades thin, broadly lanceolate to oblong, 2–8.5 cm long, 1–3 cm wide, acute to subacute or obtuse at the apex, sagittate to deeply cordate with small rounded or triangular lobes at the base, ciliate or eciliate at the margin; upper surfaces green, glabrous to densely pubescent with appressed simple multiseriate hairs to 1 mm long; lower surfaces light green, glabrous; midveins beneath usually retrorsely prickly near the base; petioles 0.5–4 cm long, glabrous, retrorsely prickly along the angles. *Ocreae* scarious, oblique, 0.5–1.3 cm long, glabrous, often weakly ciliate at the tip, lower ones often lacerate. *Inflorescences* terminal or in the axils of upper leaves, elongate, often slender, 3–15 cm long, usually branched, usually glabrous but sometimes bearing a few recurved prickles in lower portion; inflorescence bracts resembling ocreae but much reduced, 1.5–3 mm long; each inflorescence branch terminated by a dense headlike cluster of many flower fascicles. *Flower fascicles* usually 2- to 3-flowered, closely spaced; bracts lanceolate to narrowly elliptic, 3–5 mm long, glabrous, strongly acuminate; bracteoles narrowly elliptic to elliptic, 3–4 mm long, glabrous, often ciliate at the tip; pedicels short, 1–1.5 mm long, glabrous, included in bracteoles; perianth 5-parted, white to greenish-white, often reddish, 3–5 mm long; tepals broadly elliptic, obtuse, not becoming fleshy; stamens eight, inserted at the base of perianth in two whorls; inner stamens three, ca. 2 mm long; outer stamens five, 1–1.5 mm long; style one, 3-cleft to middle, ca. 0.5 mm long, included; stigmas three. *Achenes* sharply trigonous, usually ovoid in outline, acute at the apex, dark brown to black, punctuate to smooth, dull to shiny, 3–4 mm long.

Distribution. In moist, grassy situations; often in partial shade; occurring naturally in meadows, pastures, margins of swamps, ponds, and along stream banks; common as a weed along roadside and in other disturbed sites; in Asia, from northern India through China, Manchuria, and eastern Siberia to Korea and Japan (Fig. 30); in eastern North America, from Manitoba to Newfoundland, southward to Florida and Texas (Fig. 31).

Polygonum sagittatum is an extremely variable species and known both from North America and Asia. Some authors (Hara, 1952; Meisner, 1856; Ohki, 1926; Ohwi, 1965) pointed out on the basis of differences in achene surface and leaf margin that North American populations are separable from Asiatic ones, and treated them as two distinct varieties of *P. sagittatum* (Hara, 1952), or as two distinct species, *P. sagittatum* in North America and *P. sieboldii* in eastern Asia (Meisner, 1856; Ohki, 1926; Ohwi, 1965). The distinguishing characters show weak geographical variation; the following forms can be recognized.

(1) Populations from central and southern Japan, Korea, and part of China have glabrous, completely eciliate leaves. Achene surfaces in these populations are highly variable, from punctuate and dull to smooth and shiny. This form has been called *P. sieboldii.* Plants of this form with completely smooth and shiny achenes also occur sporadically in North America and was described as *P. sagittatum* var. *ovalifolium.*

(2) Plants from the eastern United States and southern Canada have glabrous, strongly ciliate leaves. Achene surfaces are always smooth and shiny. This form has been called *P. sagittatum.*

(3) A high proportion of plants from northern India, Yunnan, eastern Siberia, northern Japan, and northeastern United States and Canada has weakly to densely pubescent leaves, frequently with ciliate margins. Achene surfaces are completely smooth and shiny. This form also occurs sporadically in the southern United States. It has been called *P. sagittatum* var. *sibiricum* in Asia.

As Steward (1930) pointed out, North American populations are well within the range of variability of eastern Asiatic populations and both regions basically show similar variations in achene surfaces and leaf margin. In addition, populations from both regions did not show any

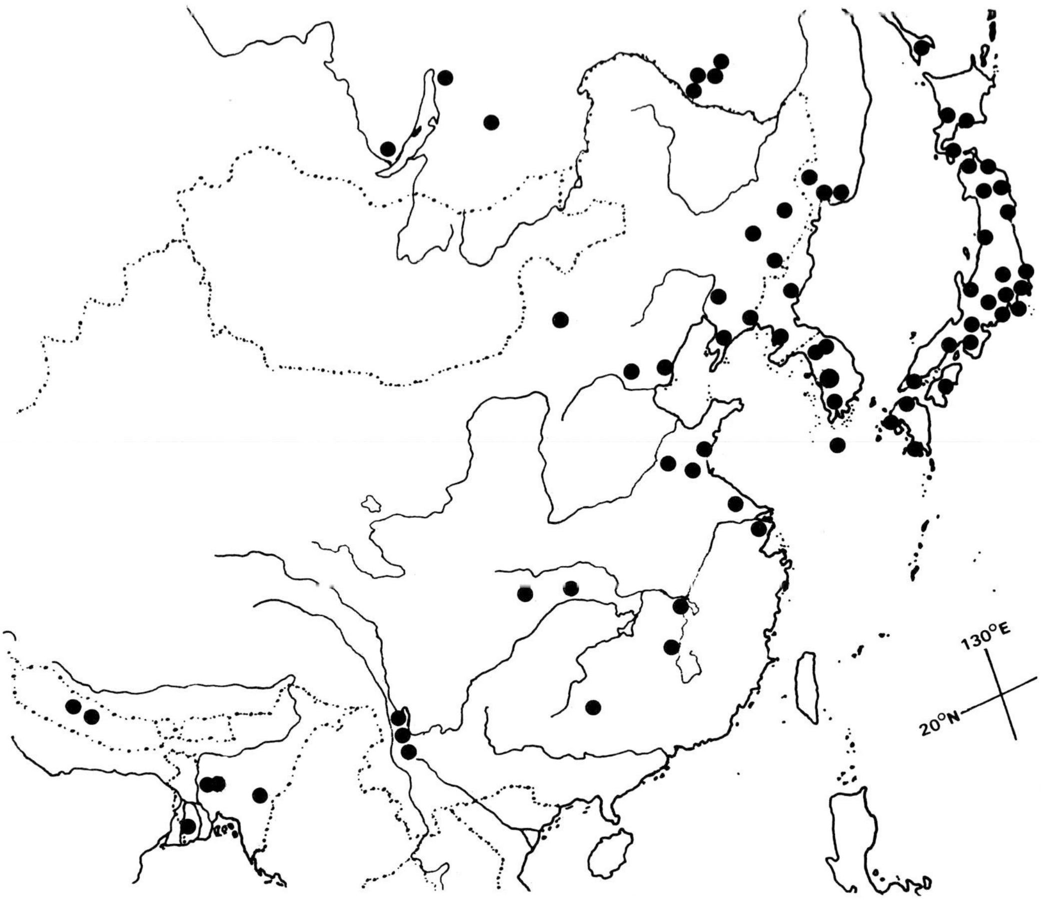

Fig. 30. Distribution of *P. sagittatum* in Asia.

substantial qualitative differences in flavonoid profiles (Park, 1987). Therefore, no infraspecific taxa are recognized.

There are other names that apparently belong to *P. sagittatum,* but I am unable to determine their identities, because the types have not been located. The list of these names is as follows:

(1) *Polygonum sagittatum* var. *paludosum* Komarov, Trudy Imp. S.-Peterburgsk. Bot. Sada **22:**132. 1903.
(2) *Polygonum sagittatum* var. *sericeum* Nakai, J. Coll. Sci. Imp. Univ. Tokyo **23:** 16. 1908.
(3) *Persicaria aestiva* Ohki, Bot. Mag. (Tokyo) **40:** 56. 1926.
(4) *Persicaria sieboldii* var. *tomentosa* Hara, J. Jap. Bot. **9:** 125. 1933.
(5) *Persicaria sieboldii* var. *brevifolia* Kitamura, Rep. 1st Exp. Manch. **4(4):** 76. 1936.

Additional specimens examined. It seems unnecessary to cite the thousands of specimens collected from North America on deposit in herbaria across the country, thus only Asian collections are listed. BANGLADESH. East Bengal, 1863–1864, *Griffith 4125* (K).

CHINA. Northern China, 24 Aug 1922, *Licent s.n.* (K, P). CHEKIANG: Ningpo, Aug 1866, *Faber s.n.* (K). HEILUNGKIANG: Lake Khanka, 1898–1899, *Bohnhof s.n.* (K, P). HOPEH: Chihli Prov., 18 Oct 1918, *Clemens 8053a* (E); Peitaiho, 20 Jun 1930, *Hsia 1909* (NY); Changli, 21 Jun 1930, *Hsia 1939* (NY); Hsiao Ling Men, Aug 1936, *Liu 12677* (GH, NY). HUNAN: Sinning Hsien, Ma-Ling-Tung, 650 m, 23 Sep 1935, *Fan & Li 535* (GH, L). HUPEH: Without specific locality, 1885–1888, *Henry s.n.* (GH, US); Ichang, Mar 1886, *Henry 579* (K); Ichang, Feb 1887, *Henry 2514* (K); Ichang, Feb 1887, *Henry 3172* (K); without specific locality and date, *Wilson 2630* (NY). INNER MONGOLIA: Jehol, without date, *David 2184* (P). KIANGSI: Lushan, in 1975, *Chow 145* (E, K, MO); Lushan, from Hwang-lung-ssu to Chiao-luh-chiao, 970–990 m, 30 Aug 1932, *Tsiang 10707* (NY). KIANGSU: Haichow, 22 Oct 1932, *Chang & Cheng 904* (NSM); Liayun Gang, 9 Sep 1981, *Liu*

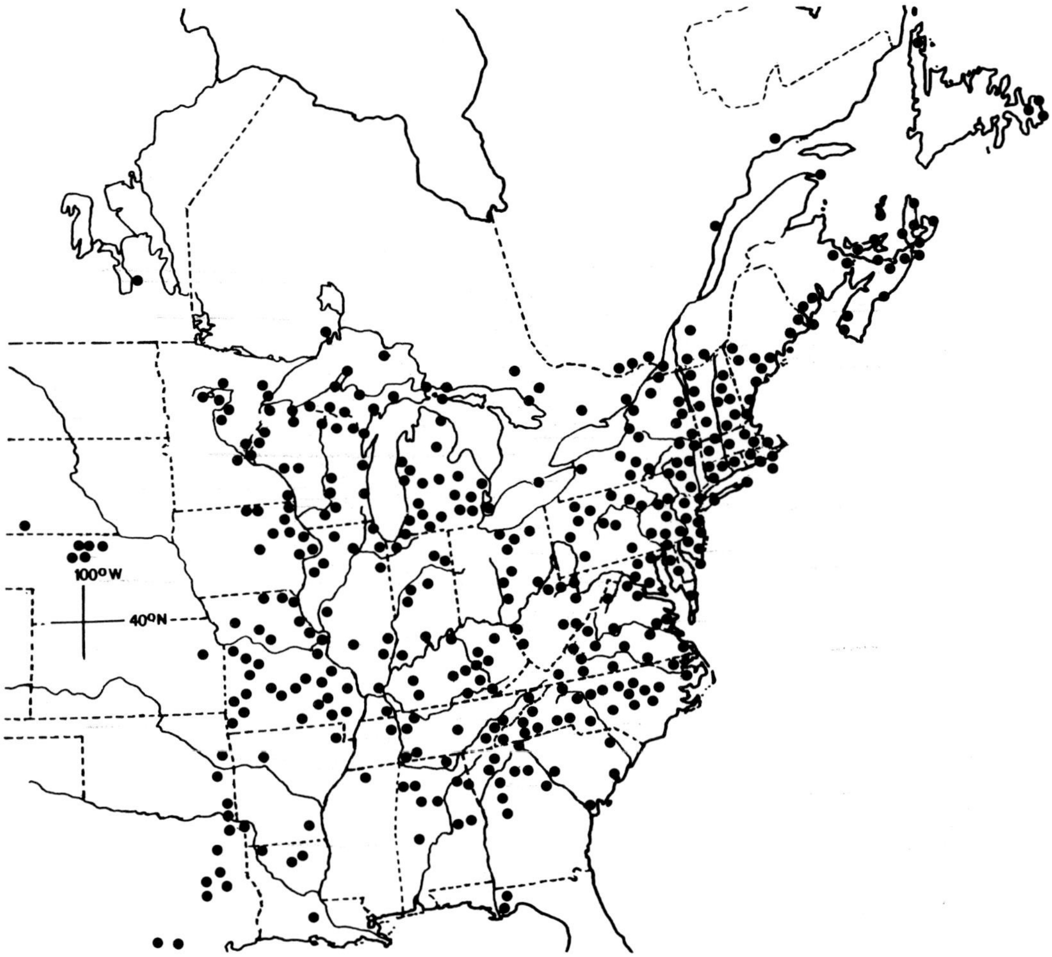

Fig. 31. Distribution of *P. sagittatum* in North America.

& *Yao 8378* (NSM). KIRIN: Without specific locality, 3 Aug 1896, *Komarov s.n.* (P); Changbaishan, on edge of forest, 800 m, in 1981, *Wan & Chow 81091* (K). LIAONING: Liao-tung Peninsula, near Sta. Valfangoo, 20 Jul 1902, *Litvinov s.n.* (NY); nr. Sta. Shitorheza, 30 Jul 1902, *Litvinov s.n.* (NY); Antung, Jul 1928, *Makino 16775* (MAK); nr. Mukden, in 1887, *Webster s.n* (K); without specific locality, 8 Sep 1959, *Col.?* (PE). PEKING: Without specific locality, in 1960, *Chow 3740* (A); without specific locality, in 1869, *David s.n.* (P). SHANTUNG: Fei Hsien, Meng Shan, 12 Jul 1936, *Cheo & Yen 37* (GH); Taishan, Wuchimiao to Wuhsientze, 24 Sep 1935, *Wang 555* (K). SZECHWAN: Tchen-keou-tin Dist., without date, *Farges 154* (NY, P, US). YUNNAN: Without specific locality and date, *Delavay 3209* (P); without specific locality, 8 May 1900, *Ducloux 625* (K); hills around Ta-chu-pa, Aug 1924, *Forrest 24845* (E, K,US); without specific locality, 10 Aug 1921, *Maire 266* (P); mountain pasture nr. Tengchuan, 2900 m, without date, *Maire 6446* (GH, LE); "marshes of Ié-nur-tch'oan," 3200 m, without date, *Maire 6371* (GH, US); without specific locality and date, *Maire 6466 & 7481* (US).

INDIA. ASSAM: Khasi Hills, Mawphlang, 6000 ft, Aug 1954, *Chand 8049a* (L, MICH); Mt. Khasia, 5–7000 ft, without date, *Hook. f. & Thomson s.n.* (GH); Khasi Hills, Mawryngkneng, 4000 ft, 11 Sep 1951, *Koelz 28437* (MICH). MANIPUR: Karong, ca. 3500 ft, 27 Sep 1950, *Koelz 26300* (L, MICH); Myang Khong Valley, 4000 ft, 20 May 1882, *Watt 7149* (K, P). PUNJAB: Manali, Kulu, 9 Oct 1916, *Cooper 5611* (MICH); Kulu, Sep 1853, *Edgeworth 29* (K); Kangra Dist., in 1921, *Sahni s.n.* (K).

JAPAN. Without specific locality and date, *Faurie s.n.* (NY). AKITA: Mt. Taihei-zan, 9 Oct 1980, *Endo 896* (TUS). AOMORI: Sanbongi, in 1886, *Faurie 43* (MO, P); Kamikita-gun, Nohezi-cho, 7 Sep 1910, *Makino 16777* (MAK). CHIBA: Ichikawa-shi, Mama, 6 Oct 1895, *Makino 16791* (MAK); Isumi-gun, Otaki-cho, Oct 1904, *Makino 17208* (MAK); Nagao-gun, Nagaogura, Iwanuma, 2 Jun 1912, *Makino 17264* (MAK); Shimousa, Higashikatsushika-gun, 23 Oct 1966, *Ohba et al. 279* (FLA). **Fukuoka:** Kasuya Co., Tatara, 1 Jul 1924, *Tanaka 100277* (US). **Gifu:** Gifu-shi, Nagaragawa, nr. the bridge of Nagara-bashi, 20 m, 6 Oct 1981, *Endo 423* (TUS); Prov. Mino, 5 Oct 1921, *Shiota 3785* (GH); 16 Apr 1929, *3786* (GH); 7 Jun 1929, *3873* (GH); 13 Oct 1930, *4381* (GH); 29 Aug 1934, *7859* (GH); 14 Jun 1936, *9754* (GH). **Gunma:** Azuma-gun, Takayama-mura, 6 Sep 1964, *Matsuda s.n.* (MAK 60106); Azuma-gun, Naganohara-cho, Kitakaruizawa, 27 Sep 1966, *Matsuda s.n.* (MAK 65162); Maebashishi, 1 Jun 1906, *Mishiro s.n.* (MAK 17259); Kurotakiyama, 23 Sep 1956, *Satomi s.n.* (MAK 1853). **Hokkaido:** Sapporo, 4 Sep 1903, *Arimoto s.n.* (GH); Mombetsu, 17 Sep 1887, *Faurie 1158* (MO); without specific locality, Sep 1904, *Faurie 5853* (BM); Sapporo, 4 Sep 1915, *Kudo s.n.* (TUS 16954); Prov. Iburi, Tomakomai, 30 Sep 1919, *Kudo s.n.* (TUS 16890); Hakodate, in 1861, *Maximowicz s.n.* (GH, K). **Ibaragi:** Daigomachi, Mt. Yamizo-san, 700–1000 m, 7 Oct 1983, *Nagayama 376* (TUS). **Iwate:** Esashi-shi, Ite-mura, 9 Oct 1904, *Kagawa s.n.* (MAK 16782). **Kagoshima:** Prov. Satsuma, 19 May 1960, *Sugaya s.n.* (TUS 16957). **Kanagawa:** Kawashima, 18 Jun 1915, *Makino 17291* (DAO); Kawasaki City, Suge, 11 Nov 1923, *Makino 17218* (DAO); Yokohama, in 1862, *Maximowicz s.n.* (K, L, NY, US); Yokoska, 1866–1874, *Savatier 1037* (K, P, US); Hakone, Aug 1871, *Savatier s.n.* (A, P). **Kochi:** Aki-shi, Doi, 3 Jun 1892, *Makino 17270* (MAK); Takaoka-gun, 13 May 1905, *Makino 17269* (MAK); Agawa-gun, Gohoku-mura, Jingamori, 30 Oct 1960, *Yamanaka s.n.* (TUS). **Kyoto:** Prov. Yamashiro, Tanabe-cho, Tsuzuki-gun, 6 Jun 1965, *Murata 19592* (L, MSC, NY); Mt. Jyubu, Wazuka-cho, Soraku-gun, 300–600 m, 17 Sep 1978, *Tsugaru 5100* (MO). **Mie:** Ueno-shi, Jun 1904, *Koreishi s.n.* (MAK 17267). **Miyagi:** Shichigashuku-machi, 4 Sep 1978, *Aiba s.n.* (TUS 67983); Rikuzen, Sendai-shi, 7 Sep 1913, *Endo s.n.* (TUS 16951); Sendai, Dainenji-yama, 2 Oct 1931, *Kimura & Naohara s.n.* (TUS 16946); Rikuzen, Miyatojima, Ootakamori, 23 Sep 1957, *Kyogaku s.n.* (TUS 16958); Kami-gun, Nakaniida-machi, Naruse River, 23 Sep 1978, *Naito & Ishikawa s.n.* (MO, TUS); Rikuzen, Okunikkawa, 2 Oct 1960, *Ohashi s.n.* (TUS 50970); Rikuzen, Sendai-shi, Sabo-yama, 8 Oct 1980, *Suzuki 94* (TUS). **Nagano:** Minami-karuizawa, 13 Sep 1954, *Koyama s.n.* (DAO); Minamisaku-gun, Koumi-machi, Ikenotaira, Mt. Yatsugatake, 1500–1600 m, 12 Aug 1978, *Murata & Ohba 5375* (TUS). **Nagasaki:** Nishimatsuura-gun, Nishiarita-cho, 15 Oct 1905, *Chiba s.n.* (MAK 16815); Isl. Uku-shima near Taira, 29 Sep 1979, *Tateishi 4660* (TUS). **Osaka:** Prov. Idzumi, Kishiwada, Mt. Konoyama, 12 Oct 1972, *Oya s.n.* (E). **Saitama:** Prov. Musashi, 13 Oct 1926, *Masamune s.n.* (NY); Koshigaya in Musashi, 24 Sep 1950, *Ohwi 110* (E, F, L, MO, MT, NY, US); Musashi, Wadamura, 30 Oct 1893, *Watanabe s.n.* (GH, US); Musashi, Ogiruba?, 24 Oct 1908, *Col.?* (LE); Prov. Musashi, Itabashi, 20 Sep 1910, *Col.?* (LE). **Shiga:** Kohga-gun, Shigaraki-machi, Miyamachi, SW foot of Mt. Handosan, ca. 300 m, 16 Sep 1982, *Ohashi et al. 9110* (TUS). **Shizuoka:** Fuji-gun, Shibakawa-cho, without date, *Kato s.n.* (MAK 16805). **Tochigi:** Nikko, Numanotaira, 1300 m, 22 Sep 1966, *Kanai 10383* (F, FLA, K, MAK, P); Nikko, Jul 1926, *Makino 24095* (L); Nikko, 4 Aug 1928, *Takeda s.n.* (E, K). **Tokyo:** Suginami-ku, Wadamura, 3 Jun 1888, *Makino 17263* (MAK); Shibuyaku, Onden, Jun 1900, *Makino 17290* (L, MAK); Itabashi-ku, 29 Oct 1904, *Makino 16821* (DAO, L); Shibuya-ku, Shibuya, 12 Oct 1923, *Makino 16820* (DAO, MAK); without specific locality, 30 Sep 1897, *Matsumura s.n.* (US); without specific locality, Aug 1905, *Terasaki s.n.* (K); nr. Tokyo, Oct 1909, *Col.?* (E). **Toyama:** Toyama-shi, Kurehayama, 14 Oct 1928, *Matsumoto s.n.* (MAK 17212). **Yamagata:** Nr. Atsumi Village, 25 Sep 1951, *Walker 7339* (US). **Yamaguchi:** Iwakuni City, Village of Atago, 19 Sep 1953, *Charette 1408* (MO); Yanai-shi, Sanai-mura, 6 Jun 1905, *Makino 17268* (MAK); Abu-gun, Ato-cho, Tokusa, 23 Sep 1968, *Miake 180* (TUS). **Yamanashi:** Minamikoma-gun, Kushigata-yama, nr. Himuro Shrine, 13 Oct 1983, *Endo 824* (TUS); Prov. Kai, Kitatsuru-gun, Uenohara-machi, in 1932, *Makino 17282* (DAO).

KOREA. Without specific locality, 26 Jul 1949, *Chung 4125* (MICH). **Cheju:** "Quelpaert," 20 Sep 1908, *Taquet 1287* (K). **Chunnam:** Mt. Chiri, 20 Aug 1934, *Smith s.n.* (US). **Hamnam:** Kaema Plateau, 22–24 Aug 1932, *Kitamura s.n.* (MICH); Sam-bang, 16 Sep 1934, *To 1596* (SNU). **Kangwon:** Mt. Dae-am, 22 Sep 1967, *Lee & Lee 630* (US); Yong-pyung, Mt. Balwang, wet place nr. ditch, 21 Aug 1981, *Park 580* (BH); Hyang-ro-bong, 30 Jul 1961, *Yang s.n.* (MAK 44609). **Kyunggi:** Mt. Sam-sung, 13 Sep 1938, *Chung s.n.* (SNU); Kwang-reung, 29 Jun 1944, *Chung s.n.* (SNU); Osan, 25 Aug 1969, *Migo s.n.* (TUS); Kwang-reung, 4 Sep 1949, *Oh 7346* (MICH); Mt. Chun-ma, wet place along small stream, 20 Jul 1981, *Park 540* (BH); Kwang-reung, wet place along roadside, 22 Jul 1981, *Park 550* (BH); Mt. Chu-kum, wet place along ditch, 5 Aug 1981, *Park 920* (BH); Chung-pyung, wet, shady place along roadside, 12 Aug 1981, *Park 942* (BH). **Pyungnam:** Chinnampo, Aug 1906, *Faurie 973* (BM). **Seoul:** Taeneung, 3 Sep 1942, *Chung s.n.* (SNU); without specific locality, Sep 1906, *Dunn 4300* (K); without specific locality, 10 Sep 1933, *Smith s.n.* (US).

NEPAL. Between Gargiankot and Munigaon, SE of Jumla, 10,500 ft, 22 Jul 1952, *Polunin et al. 4842* (E); Dhorpatan, 9500 ft, 7 Nov 1954, *Stainton et al. 3450* (A).

U.S.S.R. **Amur:** Prov. Transbaicalia, Dist. Bargusinsk, 9 Aug 1916, *Kanevsky 2839* (K); Blagowjestschensk, Jul 1898, *Karo 147* (BM, K, P); Blagowjestschensk, in 1906, *Karo 2188* (E, K); "Amur medius," in 1891, *Korshinsky s.n.* (GH, US). **Chita:** Nerschinsk, in 1892, *Karo 315* (BM, K, MO); Dahuria, without date, *Sosnin s.n.* (NY). **Maritime Territory:** Ussuri, without date, *Maack s.n.* (P); Vladivostok and vic., May–Oct 1919, *Topping 2483* (US). **Sakhalin:** Korsakov, Aug 1908, *Faurie 636* (P).

18. Polygonum meisnerianum Cham. & Schldl., Linnaea **3:** 40. 1828. *Persicaria meisneriana* (Cham. & Schldl.) Gómez Maza, Anales Inst.

Segunda Enseñanza Habana **2:** 278. 1896. *Tracaulon meisnerianum* (Cham. & Schldl.) Greene, Leafl. Bot. Observ. Crit. **1:** 22. 1904. *Truellum meisnerianum* (Cham. & Schldl.) Sojak, Preslia **46:** 146. 1974. Type. Brazil. "E. Brasilia," *Sellow s.n.* [holotype, HAL (not seen); one sheet believed to be an isotype at K!].

Polygonum chamissoeanum Weddell, Ann. Sci. Nat. Bot., sér. 3. **13:** 254. 1849. Type. Brazil. "In paludibus turfosis prope Barbacena in provincia Minas-Geraës," without date, *Weddell 1185* (holotype, P!).

Polygonum refractum Martius ex Meisner *in* DC, Prodr. **14:** 132. 1856, pro syn.

Polygonum meisnerianum var. *setosum* Chodat & Hassler, Bull. Herb. Boissier, sér. 2. **3:** 392. 1903. Type. Paraguay. "In regione collium, Cerros de Tobaty," Sep 1900, *Hassler 6451* (holotype, K!).

Erect to scandent *herb* arising from a fibrous root system, 0.3–1 m tall; stem branched, angled, retrorsely prickly at least in upper portion of internodes, often pubescent with minute 2- to 4-rayed tufted hairs and glandular hairs; glandular hairs stalked, 0.5–1 mm long. *Leaves* subsessile to short-petioled; blades linear lanceolate to narrowly elliptic, 3.5–14 cm long, 0.8–2 cm wide, acuminate at the apex, weakly sagittate to cordate at the base, ciliate at the margin; both surfaces glabrous to densely pubescent with appressed simple multiseriate hairs to 1 mm long, often mixed with some glandular hairs; midveins and major veins beneath sparsely to moderately retrorsely prickly. *Ocreae* scarious, cylindrical, 0.8–2 cm long, obliquely opened, lower ones often lacerate, bearing small recurved prickles at the base. *Inflorescences* terminal or in the axils of upper leaves, elongate, 5.5–14.5 cm long, branched 1–3 times, often geminate, moderately to densely glandular-pubescent at least in upper portion, often bearing bristlelike prickles below; lower inflorescence bracts resembling ocreae, 3–6 mm long; upper ones much reduced, 1.5–3 mm long, somewhat funnelform, strongly ciliate at the margin; each inflorescence branch terminated by an interrupted panicle or a somewhat headlike cluster of 3–10 flower fascicles. *Flower fascicles* 2- to 3-flowered; bracts broadly elliptic, 2–2.5 mm long, ciliate at the tip, often weakly to moderately hispid along the veins, rarely bearing a few glandular hairs along the margin; bracteoles broadly elliptic, 1.5–2 mm long, glabrous, somewhat ciliate at the tip; pedicels short, 2–3.5 mm long, almost included, often slightly exserted at maturity; perianth 5-parted, white to pink, 2.3–3 mm long; tepals broadly elliptic to ovate, not becoming fleshy; stamens five, inserted at the base of perianth in one whorl, ca. 1 mm long; style one, 3-cleft up to middle, 0.8–1 mm long, included; stigmas three. *Achenes* trigonous, dark brown to black, smooth, shiny, ovoid to obovoid in outline, acute at the apex.

The two varieties of *P. meisnerianum* can be distinguished by characters of leaf shape and ocrea. Flavonoid profiles of these varieties did not show any significant qualitative differences (Park, 1987), supporting the placement of *P. beyrichianum* as a variety of *P. meisnerianum* (Meisner, 1855). The two varieties may be distinguished as follows:

1. Leaves petioled, lanceolate to narrowly elliptic with sagittate to weakly hastate bases; ocreae somewhat truncate at the tip, usually pubescent with bristlelike hairs along the veins. var. *meisnerianum.*
1. Leaves subsessile, rarely short-petioled, linear lanceolate to lanceolate with weakly cordate bases; ocreae oblique at the tip, glabrous, rarely glandular-pubescent. var. *beyrichianum.*

18a. P. meisnerianum Cham. & Schldl. var. **meisnerianum.**

Stem moderately to densely armed with rather bristlelike recurved prickles to 2.5 mm long, often moderately to densely pubescent with varying densities of minute 2- to 4-rayed tufted hairs and glandular hairs. *Leaves* short-petioled, upper ones often subsessile; blades lanceolate to narrowly elliptic, 3.5–11.5 cm long, 1–2 cm wide, usually weakly sagittate or weakly hastate with small rounded lobes at the base; upper surfaces sparsely to densely pubescent with appressed simple multiseriate hairs, often mixed with some glandular hairs; lower surfaces almost glabrous to moderately pubescent with appressed simple multiseriate hairs, often with a few glandular hairs; midveins and major veins beneath often bearing tufted hairs and glandular hairs toward the base; petioles very short, 0.3–1 cm long, densely hispid. *Ocreae* somewhat truncate and shortly ciliate at the tip, usually moderately to densely pubescent with bristlelike hairs along the veins; pedicels usually glabrous, rarely bearing a few glandular hairs.

Distribution. In southeastern South America (eastern Brazil, northeastern Argentina, and Paraguay; Fig. 32).

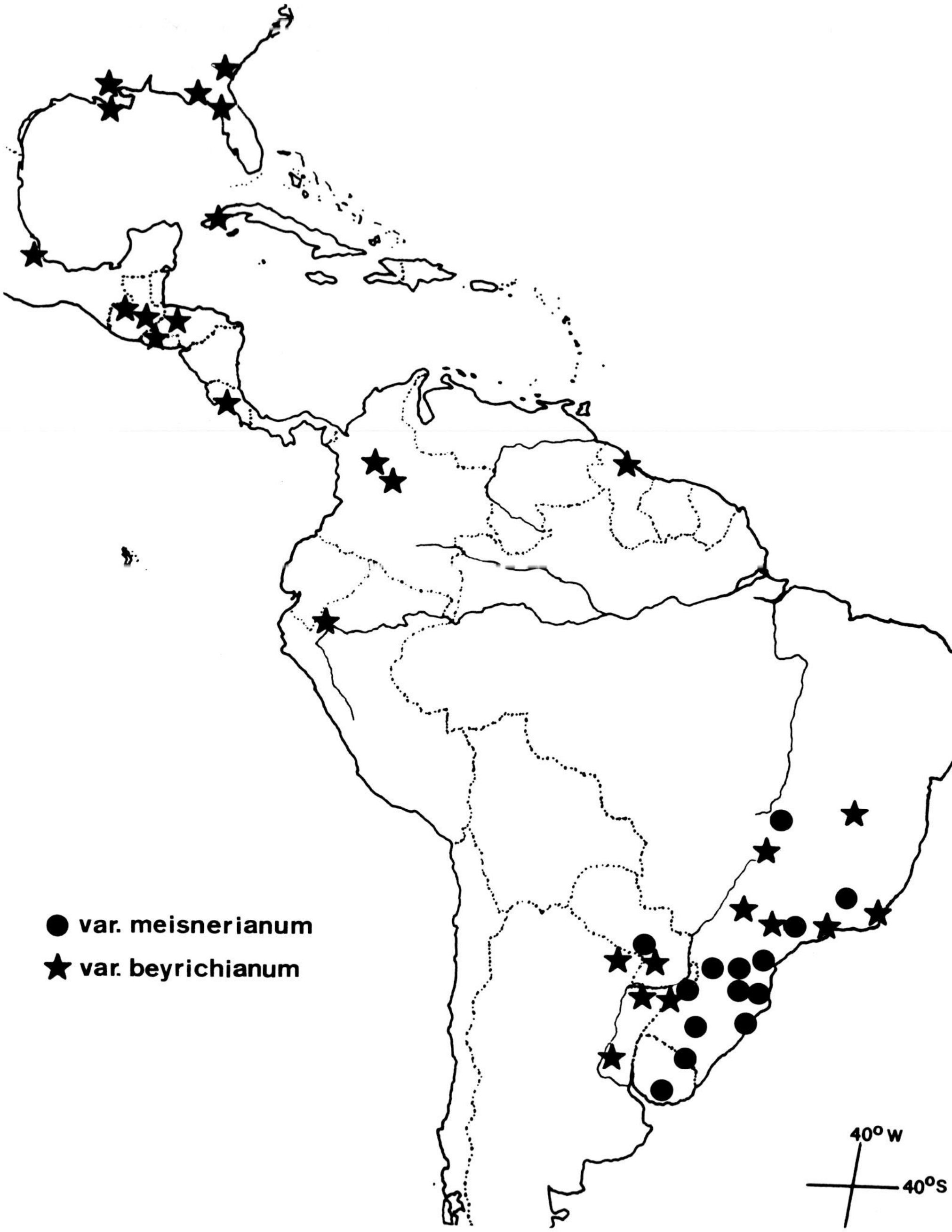

Fig. 32. Distribution of *P. meisnerianum* in North and South America.

Additional specimens examined. ARGENTINA. CORRIENTES: Posadas, 12 Mar 1907, *Ekman 2023* (F, US); Dept. Santo Tomé, Establecimiento Las Marias, Ruta Nac. No. 14, 7 km S de Gdor. Virasoro, 2 Dec 1970, *Krapovickas et al. 16914* (NY).

BRAZIL. **BRASILIA, F.D.:** Gallery Forest, Jardim Zoolôgico, 975 m, 18 May 1966, *Irwin et al. 15887* (F, NY); without specific locality, 12 Sep 1964, *Irwin & Soderstrom 6132* (F, NY); Parque do Guará, 15 Feb 1968, *Philcox & Onishi 4344* (K, NY); vic. Sobredinho, marshy ground by stream, 19 Sep 1964, *Prance & Silva 59081* (F, K, MO, NY, US). MINAS GERAIS: Caldas, 18 Mar 1866, *Regnell II-233* (NY, P, US); without specific locality and date, *Regnell s.n.* (P, US); without specific locality, in 1845, *Widgren s.n.* (K, P). PARANÁ: Curitiba, in paludosis, 28 Dec 1913, *Dusen 14336* (MICH, MO, US); Pinha, 27 May 1951, *Hatschbach 2270* (US); Mun. Curitiba, Rio Iguaçu, 28 Feb 1966, *Hatschbach 13918* (NY, US); Mun. Pirai do Sul, Nov 1968, *Hatschbach 18796* (L); Mun. Curitiba, Barigui, 2 Dec 1971, *Hatschbach 26229* (MICH, NY); Mun. Palmas, Estr. Palmas-Ponte Serrada, 5 Jul 1971, *Hatschbach et al. 28265* (US); Curitiba, 4 Dec 1914, *Jonsson 175a* (NY); Mun. Curitiba, Cidade Industrial, 18 Dec 1980, *Kummrow 1414* (MO, UEC); Guarapuava, 14 Jul 1965, *Reitz & Klein 17640* (NY, US). RIO GRANDE DO SUL: Tupancireta, 29 Jan 1942, *Rambo 9759* (SP); Passo do Socorro, p. Vacaria, 27 Dec 1951, *Rambo 51658* (US); Passo da Guarda, p. Bom Jesus, 15 Jan 1952, *Rambo 51911* (US); Morrinhos, 2 Jul 1952, *Rambo 52163* (US). SANTA CATARINA: Cunhas, Itajaí, 4 Jan 1955, *Klein 961* (NY); Serra do Oratório, Bom Jardim, S. Joaquim, 1400 m, 19 Mar 1959, *Reitz & Klein 8666* (US); Santa Cecilia, Campo do Areão, 1200 m, 20 Apr 1962, *Reitz & Klein 12626* (L, NY, US); Valoes, 750 m, 1 May 1962, *Reitz & Klein 11556* (US); Rio Castelhanos, Cacador, 900 m, 1 Sep 1962, *Reitz & Klein 11867* (US); Mun. Curitibanos and Campos Novos, bog, W of Curitibanos toward Campos Novos, 850–950 m, 5 Dec 1956, *Smith & Klein 8296* (US); Mun. Mafra, 7–9 km W of Tinguí on the rd. to Mafra, ca. 800 m, 2 Feb 1957, *Smith & Klein 10608* (MO, US); Mun. Ponte Serrado, boggy pond margin nr. Ponte Serrado, 94 km W of Joaçaba, 700–900 m, 15 Dec 1964, *Smith & Klein 14007* (NY); Mun. Pôrto União, clay pits by Rio Iguaçu, ca. 750 m, 19 Dec 1956, *Smith & Reitz 8773* (US); Mun. Pôrto União, S of Pôrto União on the rd. to Matos Costa (42 km), bog, 750–800 m, 20 Dec 1956, *Smith & Reitz 8882* (US); Mun. Bom Retiro, Fazenda Campo dos Padres, Campo dos Padres, bog, 1650 m, 25 Jan 1957, *Smith & Reitz 10452* (US); Mun. São Joaquim, nr. Encruzilhada de Boava, 7 km S of São Joaquim, brook bank and pasture, 1300 m, 5 Jan 1965, *Smith & Reitz 14272* (NY, US); Mun. Chapecó, bog nr. Campo Erê, 900–1000 m, 29 Dec 1956, *Smith et al. 9595* (F, US). SÃO PAULO: Pirituba, 9 Feb 1914, *Brade 6735* (SP); Mun. de Cunha, 15 km SE of Cunha along rd. to Paratí, 1200 m, 18 Apr 1965, *Eiten & Eiten 6246* (K); Mun. de Guararema, marshy ground along the Rio Paratí, along the Rio–São Paulo Hwy., ca. 1 km E of boundary between mun. of Guararema and Sta. Isabel, 4 Oct 1961, *Eiten & Mimura 3367* (MO, SP); Butantan, 25 Sep 1918, *Hoehne 2416* (NY, SP); Mun. de São Paulo, ca. 10 km due S of center of São Paulo, Parque do Estado, grounds of the Instituto de Botânica, in marsh, 29 Jan 1964, *Skvortzov 273* (K, NY, US).

PARAGUAY. Without specific locality, in 1931, *Jorgensen 3740* (MO). CENTRAL: San Lorenzo, Ciudad Universitaria, without date, *Arenas 935* (MICH). MISIONES: Estancia, La Soledad, swamp, 24 Nov 1956, *Pedersen 4405* (MO, US).

URUGUAY. CERRO LARGO: Sierra de Aceguá, in 1936, *Roseugurtt B-1709* (NY). SAN JOSÉ: Barra de Sta. Lucia, Apr 1927, *Herter 914* (MO, NY, US).

18b. Polygonum meisnerianum var. **beyrichianum** (Cham. & Schldl.) Meisner *in* Martius, Fl. bras. **5(1)**: 19. 1855.

Polygonum beyrichianum Cham. & Schldl., Linnaea **3**: 42. 1828. *Tracaulon beyrichianum* (Cham. & Schldl.) Small, Fl. S.E. U.S. 380. 1903. Type. Brazil. "In paludibus prope Nov. Friburgum provinciae Rio de Janeiro Brasiliae tropicae," Dec 1822, *Beyrich s.n.* (not seen; see Park, Brittonia **38**: 399. 1986).

Polygonum multangulare Hook. & Arn., Companion Bot. Mag. **2**: 62. 1836. Type. U.S.A. Louisiana: New Orleans, in 1832, *Drummond 277* (lectotype, K!, designated by Park, Brittonia **38**: 399. 1986; isolectotype, K!).

Polygonum geniculatum Riddell ex Meisner *in* Martius, Fl. bras. **5(1)**: 20. 1855, pro syn.

Polygonum chamissonianum Steudel ex Meisner *in* Martius, ibid., pro syn.

Polygonum turkheimii Vatke ex Small, Bull. Torrey Bot. Club **19**: 370. 1892, pro syn.

Polygonum meisnerianum var. *jalapense* Robinson, Proc. Amer. Acad. Arts **35**: 324. 1900. Type. Mexico. Veracruz: In wet places nr. Jalapa, 1225 m, 5 Apr 1899, *Pringle 8111* (lectotype, CU!, designated by Park, Brittonia **38**: 399. 1986; isolectotypes, K!, MIN!, MO!, MSC!, NY!, P!, TEX!, US!); see Mitchell, Rhodora **72**: 182–188. 1970.

Stem sparsely to moderately armed with relatively short recurved prickles along the angles; prickles 1–1.5 mm long. *Leaves* subsessile, rarely short-petioled; leaf blades linear lanceolate to oblanceolate, 5.5–14 cm long, 0.8–1.5 cm wide, usually cordate, more or less clasping, rarely somewhat hastate, almost glabrous on both surfaces but rarely bearing some appressed simple multiseriate hairs on upper surface and major veins beneath, rarely densely glandular-pubescent. *Ocreae* oblique at the tip, short-ciliate, usually almost glabrous but sometimes bearing some bristlelike hairs along the veins, rarely densely glandular-pubescent; pedicels usually glandular-pubescent.

Distribution. In southeastern South America, central America, Cuba, and the southeastern United States (Florida, Louisiana, and South Carolina); also occurs in southeastern Africa (Figs. 11, 32, and 33).

Additional specimens examined. ARGENTINA. CORRIENTES: Dep. Santo Tomé, 35 km SW de Santo Tomé, Ruta Nac. 40, 5 Feb 1979, *Ahumada 3056* (F); Dep. Itati, Ruta Nac. 12, 47 km E de Itati, 26 Feb 1977, *Ahumada et al. 744* (MO); Dep. Concepción,

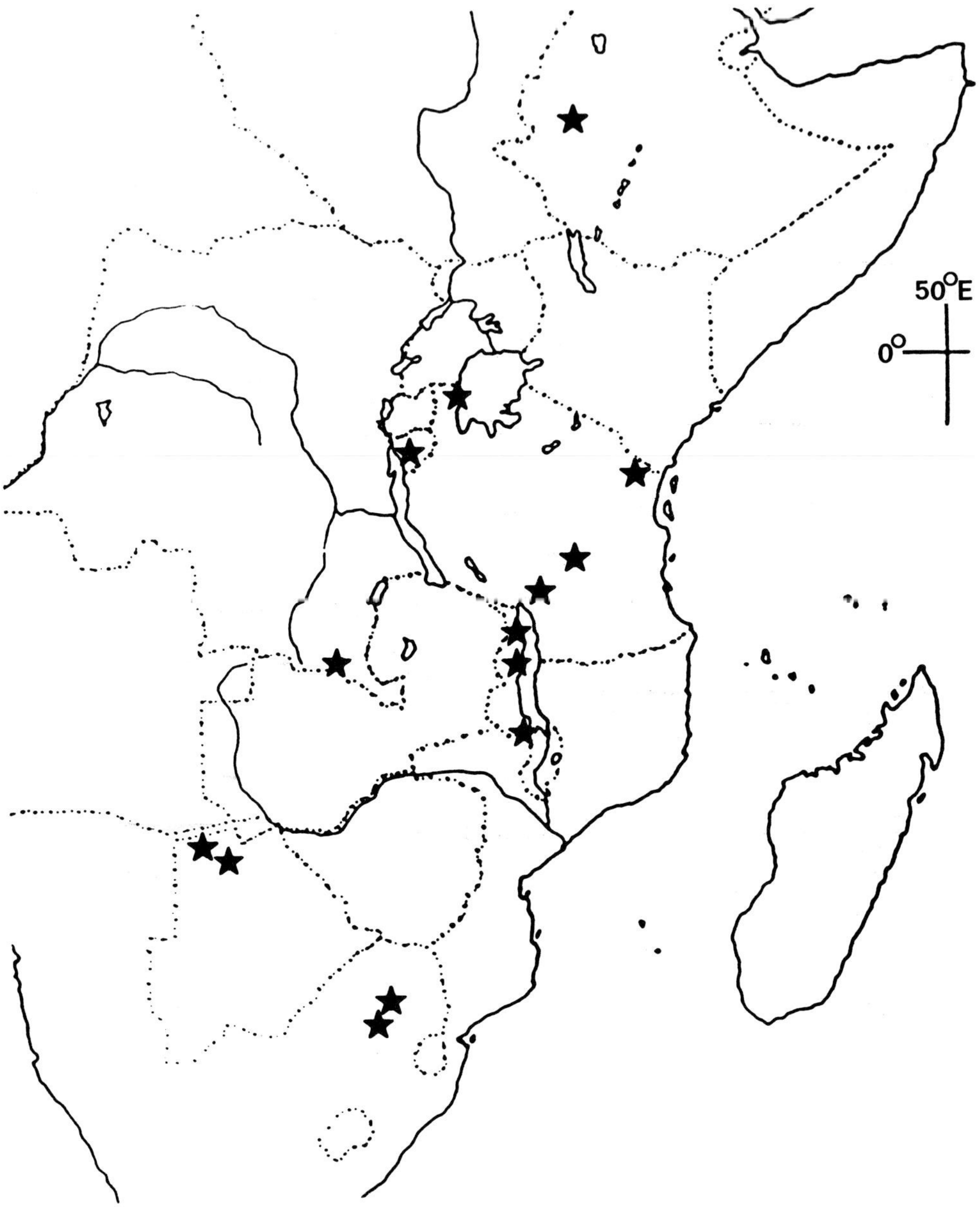

Fig. 33. Distribution of *P. meisnerianum* var. *beyrichianum* in Africa.

Estancia Millán, 6 km E de Santa Rosa, 27 Mar 1975, *Arbo et al. 815* (MICH); Dep. Santo Tomé, Ruta 40, 24 km NE de Santo Tomé, *Krapovickas et al. 25269* (MO, SI, UNC); Estancia Santa Teresa, in swamps, 26 Apr 1952, *Peterson 1698* (MO, P, US); Dep. Concepción, 28 km SE de Concepción, 21 May 1976, *Schinini et al. 13107* (SI). ENTRE RÍOS: Delta del Paraná, Arroyo Negro, 1 Mar 1936, *Burkart 7635* (K); Dep. Uruguay, Estancia La Selmira (Isla Cupalén), 31 Mar 1967, *Pedersen 8221* (K). FORMOSA: Without specific locality, Dec 1918, *Jorgensen 2170* (US).

BOTSWANA (BECHUANALAND). Khwai River, 2 hours boat trip upstream from Gobegha Lagoon, Okavango Swamp, 1000 m, 7 Mar 1972, *Russell & Biegel 1530* (MO); banks of distributionary channel of Okavango River, 26 Jan 1976, *Smith 1555* (MO); at sides of Central Boro River at 19°27¼'S, 22°56¾'E, on left bank in just over 1 m deep water—fairly open site with grasses, sedges and aquatics, 14 Apr 1977, *Smith 1964* (MO).

BRAZIL. Without specific locality and date, *Sellow s.n.* (K). MINAS GERAIS: Steep rocky lower slopes of

Pico de Itacolomi, ca. 3 km S of Ouro Preto, ca. 1650 m, 1 Feb 1971, *Irwin et al. 29565* (NY); Serra do Espinhaço, 35 km NE of Francisco Sá, road to Salinas, 1100 m, 14 Feb 1969, *Irwin et al. 23309* (F, NY); Caldas, 3 Sep 1868, *Regnell III-1207* (US). RIO DE JANEIRO: Nova Friburgo, Nov 1842, *Claussen 187* (P). SÃO PAULO: Guaratingueta, without date, *Glaziou 1144ga* (P); Mun. Moji-Guaçu, about 100 ft from the entrance of Campos das Sete Lagôas, Fazenda Campininha, just N of Rio Moji-Guaçu, 1.8 km NNW of Padua Sales (about 27 km NW of City of Moji-Mirim), 575 m, 4 Dec 1961, *Eiten 3518* (K, NY, SP, US); Bairro of Vila Maria, marshes of Rio Tietê, in shallow water and soaking soil at side of rd., 1 Apr 1960, *Eiten & Eiten 1795* (F, NY, SP, US); Mun. Iepe, Fazenda C. A. P. I., at Lagoa da Cima, 2 km due N of the Rio Paranapanema and 7 km E of Pôrto Alvorada, ca. 300 m, 9 Feb 1965, *Eiten et al. 5938* (K, US); Sorocaba, without date, *Lund s.n.* (NY).

BURUNDI. Mahwa Station de la Luvyironza, 1800 m, 31 May 1979, *Reekmans 8071* (K, MO).

COLOMBIA. ANTIOQUIA: Río Negro, 2100 m, 22 Jun 1930, *Archer 272* (MO, NY, US); San Pedro, Jun 1938, *Tomás 248* (US). CUNDINAMARCA: Barbas Co., growing in ditch, 3 Apr 1935, *Archer 3276* (US).

CONGO. Katanga Prov., 7 km S of Lukuni, 10 Dec 1970, *Symoens 13876* (K).

COSTA RICA. ALAJUELA: Laguna, Aug 1919, *Lankester K14* (K). GUANACASTE: En la Ciénaga que bordea el lago Arenal, Pcina, 500 m, 13 Dec 1964, *Jimenez 2699* (F, NY).

CUBA. Pinar del Río, in swamps along a river, 29 Oct 1923, *Ekman 17840* (NY); without specific locality, 1860–1864, *Wright 2244* (MO, NY).

EL SALVADOR. AHUACHAPÁN: Shore of Lagunita las Ninfas, Apaneca, 5430 ft, 28 Jan 1951, *Fassett 28700* (F, MO, US).

ETHIOPIA. About 5 km W of Lekemti, marshy land, ca. 1800 m, 13 Nov 1965, *Wilde 8837* (MO).

GUATEMALA. ALTA VERAPAZ: Nr. Santa Cruz, ca. 1350 m, 14 Apr 1941, *Standley 92252 & 92261* (F); large swamp E of Tactic, ca. 1450 m, 14 Apr 1941, *Standley 92507* (F); bogs nr. Santa Cruz, along rd. to San Cristobal, ca. 1350 m, 16 Apr 1941, *Standley 92760* (F); Cobán, Aug 1885, *Türckheim 243* (F, K, NY, US). JALAPA: Potrero Carrillo, at Hierba Buena, 14 mi NE of Jalapa, 1500–1900 m, 11 Dec 1939, *Steyermark 33047* (F).

GUYANA. Essequibo, Tapekuma Lake, Jun 1909, *Anderson 338* (K).

HONDURAS. INTIBUCÁ: Open forest, Cordillera Opalaca between Pela Naríz and Calaveras, rd. to La Esperanza, 3 Sep 1968, *Molina 22644* (F, MICH, NY, US); cutover cloud forest of Sierra Opalaca, above El Duraznillo, 1800 m, *Molina & Molina 26074* (F, MICH, MO, NY, US).

MALAWI. MZIMBA: Mzuzu water works, 4500 ft, 11 Feb 1974, *Pawek 8089* (MO); Champira, 5 mi NE of Katete, up Lwanjati–Musese road, 1550 m, 24 Dec 1975, *Pawek 10509* (MO). RUMPHI: Nyika Plateau, Chelinda Bridge, 7600 ft, 22 Apr 1976, *Pawek 11197* (MO).

MEXICO. VERACRUZ: Jalapa, without date, *Schiede & Deppe 89* (MO).

PARAGUAY. Without specific locality and date, *Hassler 6422* (NY); without specific locality, 24 Jan 1930, *Jorgensen 3740* (F, NY, US). CORDILLERA: Ypacaraí, in 1913, *Hassler 12502* (F, K, L, MO, NY, US). MISIONES: Estancia La Soledad, Santiago, 24 Nov 1956, *Pedersen 4405* (NY).

PERU. SAN MARTIN: Zepelacio, nr. Moyobamba, 1200–1600 m, Mar 1934, *Klug 3580* (F, K, MO, NY, US).

SOUTH AFRICA. TRANSVAAL: Belfast, 6 Feb 1904, *Davy 1319* (K); without specific locality, 1875–1880, *Rehmann s.n.* (K); Dist. Lydenburg, Jan 1896, *Wilms. 1283* (K).

TANZANIA. IRINGA: Daboga Highlands, nr. Idete School, 55 mi SE Iringa, 6000 ft, 31 Jan 1971, *Mabberley 640* (K); Nijombe Dist., Ushinga and Ubena areas, Mar 1937, *Ward 113* (K). LUSHOTO: Mkuzi, 4 mi NE of Lushoto, W. Usambaras, 20 Apr 1953, *Drummond & Hemsley 2153* (K). WEST LAKE: Bukoba, Aug 1931, *Haarer 2067* (MO).

U.S.A. FLORIDA. Alachua Co.: Gainesville, N shore of Newmans Lake nr. farm rd. with gate off route 26, growing in swampy area with water, 10 Oct 1979, *Fisher 407* (FLAS); Gainesville, Newmans Lake, 18 Jun 1935, *West s.n.* (FLAS). Jefferson Co.: Clearing of swampy woodland that forms part of hwy. right-of-way, along US Rt. 90, E of Monticello, 2.5 mi W of Jefferson–Leon Co. line, 24 Oct 1969, *Godfrey 69140* (DUKE, FLAS, FSU, LAF, TEX, UNC). Leon Co.: Scattered plants in shallow water of edge of Lake Iamoa, about 3.4 mi S of S.R. 12, 28 Jun 1977, *Wooten 3122* (FSU, LAF); in swamp clearing, opposite side of US Rt. 90 from Lake Miscosukee dam, E of Monticello, 20 Sep 1969, *Godfrey 68844* (FSU, VDB). LOUISIANA. Plaquemines Co.: Pointe á la Hache, borders of Bookfuloya River, 6 Aug 1885, *Langlois s.n.* (NY). St. Tammany Co.: Borders of Tchefuncta River, 17 Sep 1892, *Langlois s.n.* (BH, FLAS, MIN, NY); Madisonville, in 1838, *Riddell s.n.* (NY). SOUTH CAROLINA. Jasper Co.: Marsh, Savannah Wildlife Refuge nr. Hardeeville, 7 Oct 1969, *Blihoude s.n.* (UNC); wet ditch between US 601 and low pine savannah, 1.4 mi N of jct. of Co. Rt. 27–169, 22 Jun 1968, *Leonard & Radford 2231* (FLAS, FSU, GH, LAF, MICH, NLU, SP, UNC, VDB).

19. Polygonum brachypodum Baker, J. Linn. Soc., Bot. **20:** 239. 1883. Type. Malagasy. Central Madagascar, without date, *Baron s.n.* (lectotype, K!, designated by Park, Brittonia **38:** 401. 1986). Syntypes. Malagasy. Madagascar, without specific locality and date, *Hilsenberg s.n.* (K!, 2 specimens); central Madagascar, without date, *Parker s.n.* (K!).

Ascending to erect *herb,* 1–1.5 m tall; stem branched, angled, moderately to densely pubescent with 4- to 7-rayed minute stellate hairs, sparsely to moderately armed with bristlelike prickles along the angles; prickles ca. 2 mm long.

Leaves petioled, upper ones often subsessile; blades linear lanceolate to lanceolate, 5–12 cm long, 0.5–1.7 cm wide, long-ciliate at the margin, acuminate at the apex, at the base usually hastate with two acute triangular lobes but upper ones often without them; basal lobes often deflexed; upper surfaces sparsely to moderately stellate-pubescent at least along the margin, also bearing a few appressed bristlelike simple multiseriate hairs to 1 mm long; lower surfaces sparsely to densely stellate-pubescent; stellate hairs identical to those of stem; midveins and major veins beneath bearing bristlelike prickles; prickles 1–1.5 mm long; petioles 0.5–1.2 cm long, prickly. *Ocreae* scarious, 0.5–1.2 cm long, cylindrical, oblique at the apex, short-ciliate, often bearing a few bristles along the veins, often densely stellate-pubescent. *Inflorescences* terminal or in the axils of upper leaves, elongate, 7–15 cm long, branched 1–3 times, often geminate, moderately to densely pubescent with glandular hairs intermixed with varying densities of stellate hairs identical to those of stem; glandular hairs stalked, ca. 0.5 mm long; inflorescence bracts resembling ocreae but reduced, 2–5 mm long; each inflorescence branch terminated by a relatively dense panicle of flower fascicles. *Flower fascicles* 2- to 3-flowered; bracts broadly elliptic to ovate, ca. 2 mm long, densely hispid, ciliate; bracteoles elliptic to broadly elliptic, ca. 2 mm long, pubescent; pedicels 2–3 mm long, glabrous, included in bracteoles; perianth 5-parted, pink to bright red, 2–3 mm long; tepals elliptic to broadly elliptic, not becoming fleshy; stamens five, ca. 1 mm long, inserted at the base of perianth in one whorl; style one, 2-cleft, ca. 0.5 mm long, included; stigmas two. *Achenes* biconvex, ovoid in outline, 1–2 mm long, acute at the apex, smooth, shiny.

Distribution. Known only from Madagascar (Fig. 34).

Polygonum brachypodum, an endemic Madagascan species, is very poorly represented in herbaria, and it has not been possible to determine its patterns of morphological variation and relationships to other species with certainty. *Polygonum brachypodum* is probably related to *P. dichotomum* var. *angustissimum* of India. The two taxa share many characters, including ciliate, oblique ocreae, five-parted perianths, a two-cleft style, and relatively dense paniclelike inflorescences. However, *P. brachypodum* can be distinguished from *P. dichotomum* var. *angustissimum* by its hastate, stellate-pubescent leaves and smooth, shiny achenes.

Additional specimens examined. MALAGASY. MADAGASCAR: Without date and collector's name (K); central Madagascar, without date, *Baron 3710* (K); Marais, 2 Sep 1930, *DeCany 6973* (K); Nanisana, without date, *D'Alleizette s.n.* (L).

20. Polygonum dichotomum Blume, Bijdr. **11:** 529. 1826. *Persicaria dichotoma* (Blume) Masamune, Sci. Rep. Kanazawa Univ., Biol. **2:** 82. 1954. *Truellum dichotomum* (Blume) Sojak, Preslia **46:** 145. 1974. Type. Indonesia. Java: "Circa Bataviam locis humidis," *Blume s.n.* (one specimen designated as isotype by E. D. Merrill at NY, but without locality and collector's name).

Polygonum tetragonum Blume, Bijdr. **11:** 535. 1826. *Tracaulon tetragonum* (Blume) Greene, Leafl. Bot. Observ. Crit. **1:** 22. 1904. Type. Indonesia. "In altioribus montis Burangrang locis uliginosis," *Blume*? (not seen; see Meisner, Ann. Mus. Bot. Lugduno-Batavum **2:** 63. 1865).

Polygonum muricatum var. *glabratum* Wallich, Wallich's Catalogue no. 1697-3. 1829, nom. nud.

Polygonum glabratum Meisner ex Wallich, ibid. no. 6285. 1832, nom. nud.

Polygonum pedunculare Wallich ex Meisner *in* Wallich, Pl. Asiaticae Rar. **3:** 58. 1832. *Tracaulon pedunculare* (Wallich ex Meisner) Greene, Leafl. Bot. Observ. Crit. **1:** 22. 1904. *Polygonum strigosum* var. *pedunculare* (Wallich ex Meisner) Steward, Contr. Gray Herb. **5(88):** 91. 1930. *Persicaria peduncularis* (Wallich ex Meisner) Nemoto, Fl. Jap., Suppl. 174. 1936. Type. Singapore. Without specific locality, in 1822, *Wallich s.n.* (lectotype, K!, designated by Park, Brittonia **38:** 400. 1986). Syntype. Bangladesh. In Silhet, *De Silva s.n.* (not seen).

Polygonum pedunculare var. *subcordatum* Meisner *in* DC., Prodr. **14:** 134. 1856. Type. Indonesia. Java: Without specific locality and date, sent by *de Vriese* (holotype, NY!).

Polygonum pedunculare var. *latifolium* Meisner, Ann. Mus. Bot. Lugduno-Batavum **2:** 63. 1865. Type. Indonesia. Java: Without specific locality and date, *Zippelius s.n.* (lectotype, NY!, designated by Park, Brittonia **38:** 400. 1986). Syntypes. Indonesia. Java: Without specific locality and date, *Blume s.n.* (NY!). Sumatra: "Sumatra occid.," without date, *Korthals s.n.* (NY!).

Polygonum pedunculare var. *lanceolatum* Meisner, Ann. Mus. Bot. Lugduno-Batavum **2:** 64. 1865. Type. Fiji. Without specific locality, in 1860, *Seemann 370* (lectotype, NY!, designated by Park, Brittonia **38:** 400. 1986). Syntypes. Indonesia. Java, Sumatra, Borneo, *Korthals* (not seen).

Polygonum pedunculare var. *subauriculatum* Meisner, Ann. Mus. Bot. Lugduno-Batavum **2:** 64. 1865. Type.

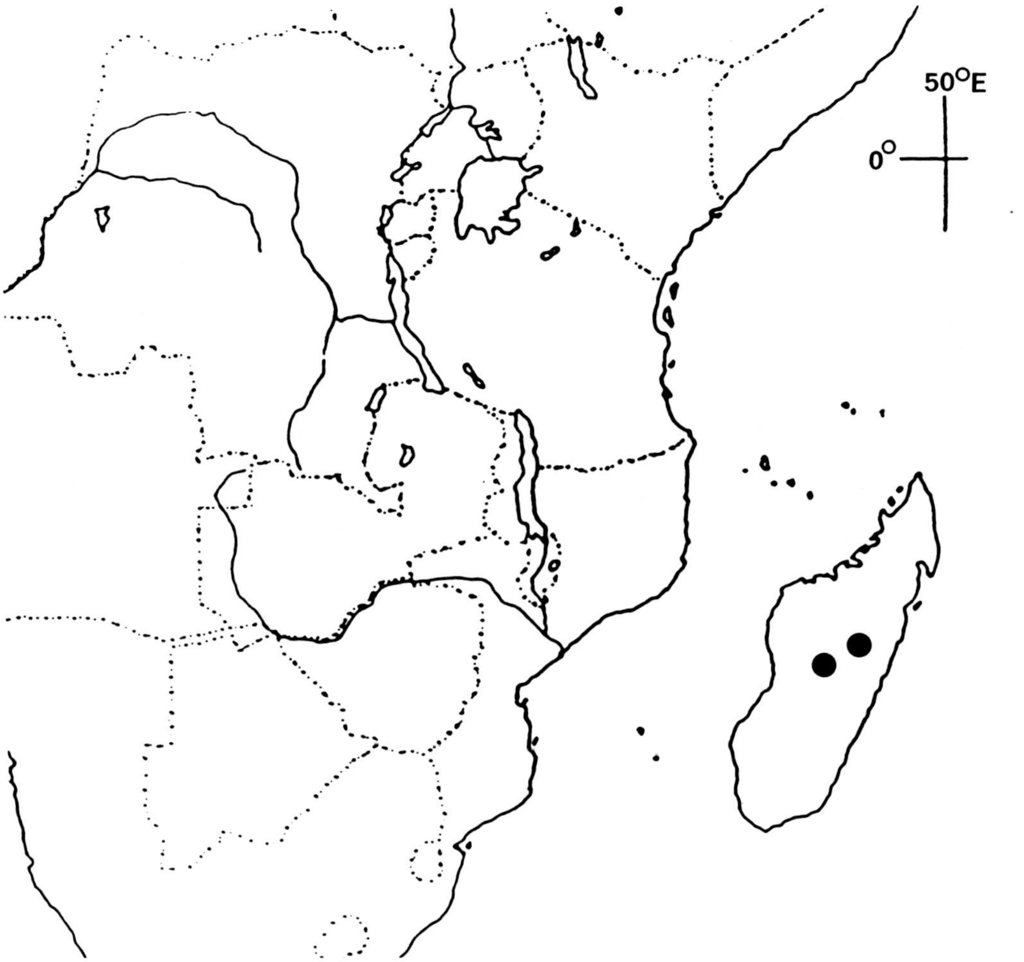

Fig. 34. Distribution of *P. brachypodum.*

Indonesia. Celebes: Gorontalo, without date, *Forsten s.n.* (lectotype, NY!, designated by Park, Brittonia **38:** 400. 1986; isolectotype, L!). Syntype. Indonesia. "Borneo australis," *Korthals s.n.* (not seen).

Polygonum javanicum Zipp. ex Meisner, Ann. Mus. Bot. Lugduno-Batavum **2:** 63. 1865, pro syn.

Polygonum imberbe Seemann ex Meisner, ibid. **2:** 64. 1865, pro syn.

Scandent to ascending *herb,* 0.3–1 m tall; stem often branched, decumbent at the base, glabrous, angled, almost unarmed to densely armed with bristlelike recurved prickles to 1.5 mm long; lower nodes often producing fibrous roots. *Leaves* petioled, often subsessile to sessile; blades lanceolate to elliptic or linear lanceolate, 5–12 cm long, 1–3.5 cm wide, acute to acuminate at the apex, at the base usually attenuate but often weakly cordate, often bearing small rounded auricules near the base, ciliate at the margin; both surfaces almost glabrous; midveins beneath often antrorsely prickly; petioles 0.5–2.5 cm long, unarmed. *Ocreae* scarious, 1.5–2.2 cm long, cylindrical, oblique at the apex, usually glabrous but sometimes bearing bristlelike hairs along the veins, rarely ciliate, lower ones often lacerate. *Inflorescences* terminal or in the axils of upper leaves, elongate, usually slender, 6–16 cm long, branched, often geminate, sparsely to densely glandular-pubescent in upper portion; glandular hairs stalked, ca. 0.5 mm long; inflorescence bracts resembling ocreae but reduced, ca. 5–7 mm long, frequently ciliate; each inflorescence branch usually terminated by a relatively dense panicle of many flower fascicles. *Flower fascicles* usually 2- to 3-flowered; bracts broadly elliptic,

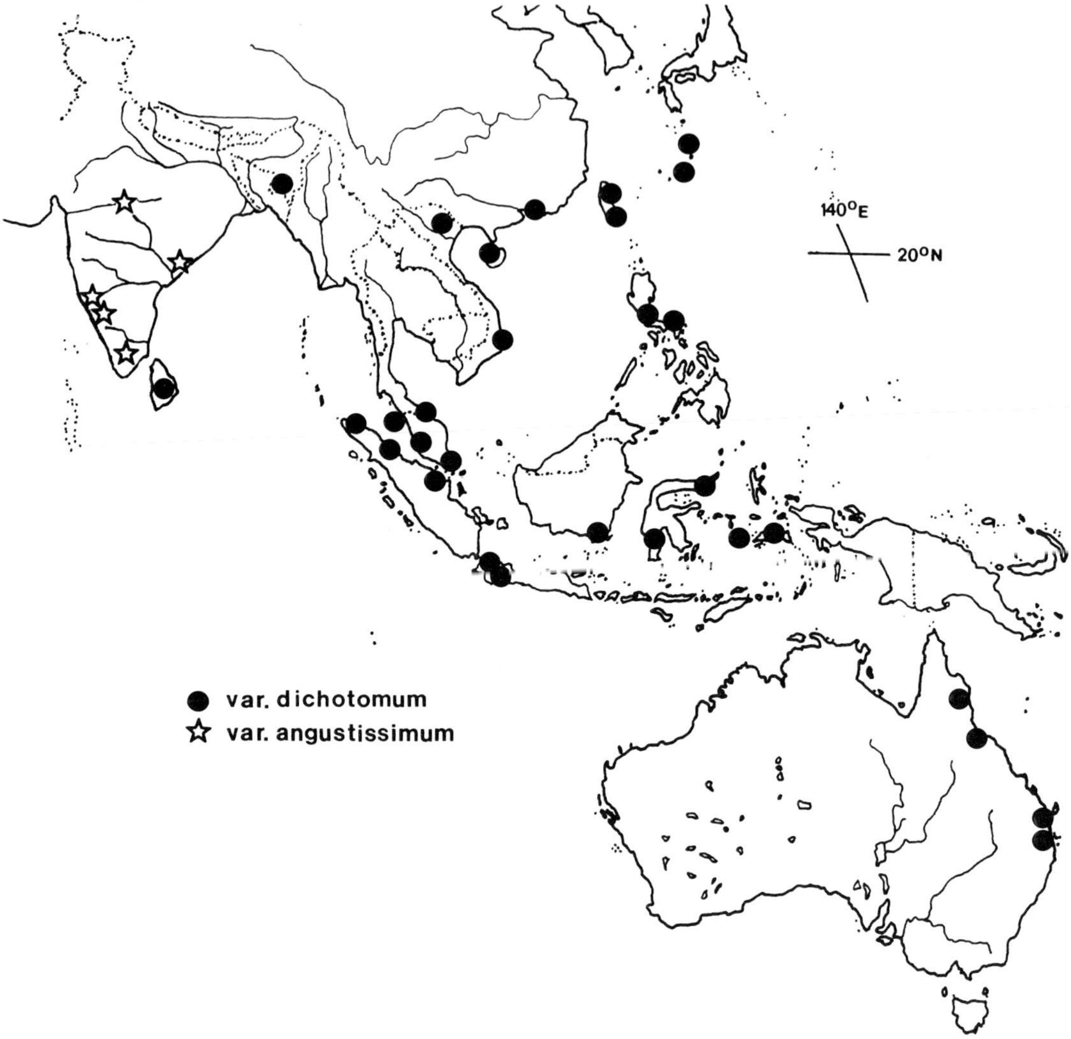

Fig. 35. Distribution of *P. dichotomum.*

ca. 3 mm long, usually glabrous, acuminate, weakly to moderately ciliate, rarely glandular-pubescent; bracteoles elliptic, 2–3 mm long, glabrous, often weakly ciliate at the tip; pedicels short, 1.5–2 mm long, glabrous, almost included in bracteoles; perianth 5-parted, white to pink, 3–4 mm long; tepals elliptic to broadly elliptic, obtuse, not becoming fleshy; stamens five, inserted at the base of perianth in one whorl, ca. 1 mm long; style one, 2-cleft, included; stigmas two. *Achenes* biconvex, ovoid in outline, acute at the apex, dark brown to black, shiny, rarely dull, 3–4 mm long.

Polygonum dichotomum displays striking variation in leaf shape, and at least eight infraspecific taxa, mainly based on leaf shape, have been described. Most specimens from southeast Asia and Australia have short-petioled, lanceolate to narrowly elliptic leaves with acute to weakly cordate bases. Petioles are 1–2 cm long, ranging from $\frac{1}{10}$ to $\frac{1}{4}$ of the length of blade. These have been called *P. pedunculare* var. *lanceolatum.* In this form, the lower leaves frequently have small rounded auricules at the base, and plants of this have been called *P. pedunculare* var. *subauriculatum.* A high proportion of collections from Java and the Philippines has long-petioled, elliptic leaves with weakly to moderately cordate bases. Petioles are relatively long, ranging from 1/5 to 1/3 of the length of blade. These plants have been called *P. pedunculare* var. *subcordatum*. However, these characters intergrade, and these varieties are not clearly delimitable entities and are interconnected by intermediate populations.

Another variant, which occurs in southern India, is rather distinct from the rest; it is recognized here as a taxonomic variety. The varieties may be distinguished as follows:

1. Leaves lanceolate to elliptic, frequently with auricules at the base, petioled; inflorescences slender, loose; achenes shiny. var. *dichotomum*.
1. Leaves usually linear lanceolate, subsessile to sessile; inflorescences stout, ascending, dense; achenes dull. var. *angustissimum*.

20a. P. dichotomum var. **dichotomum.**

Stem almost unarmed or sparsely prickly. *Leaves* petioled; blades lanceolate to elliptic, 5–8 cm long, 1–3.5 cm wide, subacute to acute at the apex, acute to weakly cordate at the base; petioles 1/10 to 1/3 of the length of blade. *Ocreae* usually glabrous but rarely bearing a few bristle-like hairs along the veins, eciliate. *Inflorescence branches* relatively slender and loose, usually sparsely glandular-pubescent; bracts weakly to moderately ciliate, glabrous. *Achenes* weakly punctuate but shiny.

Distribution. From northern India (Assam) through southern China to Ryukyu Islands, south to northeastern Australia through southeast Asia (Fig. 35).

Additional specimens examined. AUSTRALIA. QUEENSLAND: Cairns, 27 Apr 1967, *Brass 33543* (K); Palmwoods, 60 mi N of Brisbane, in swamp in cleared forest country, 150 ft, 25 May 1930, *Hubbard 2776* (L); Proserpine, without date, *Michael 851* (BM, GH); Bli Bli, ca. 12 km ENE of Nambour, 9 May 1981, *Wilson & Sharp 3817* (K).

CHINA. KWANGTUNG. Hainan: Janfengling, 800 m, in 1978, *Chow et al. 78476* (NY); without specific locality, Nov 1889, *Henry 8309* (K); Yaichow, in moist places along stream, 29 Jul 1933, *Liang 62317* (NY); Yaichow, into water or on banks in mixed woods, 25 Sep 1933, *Liang 63136* (NY); Taam Chau Dist., Sinwoh, 18 May 1928, *Tsang 371* (GH, US); without specific locality and date, *Tsang 16229* (NY, US); on ravine, 28 Jul 1933, *Wang 33418* (NY); without specific locality, 21 Aug 1933, *Wang 33821* (NY); on ravine, 16 Dec 1933, *Wang 35455* (NY).

HONG KONG. Lantac Isl., 22 Jun 1928, *Ying 730* (K, NY).

INDIA. ASSAM: Cachar, 4 Jun 1874, *Keenan s.n.* (K). MADRAS: Palni Mts., Sep 1836, *Wight 2462* (K); Nilgiri Mts., Aug 1851, *Col.*? (K).

INDONESIA. BORNEO: Marabahan, Badandan, 8 Nov 1922, *Dachlan 24* (K, L); Bangarmassing, 1857–1858, *Motley 228* (K). CELEBES: Belaang, Mongondou, Aug 1917, *Kauderns 176* (L, NY); without specific locality and date, *Kds. 18518b* (L); Pinrang, 14 Jun 1912, *Noerkas 316* (K, L). JAVA: Bantam, 90–100 m, 9 Aug 1937, *van Steenis 10478* (NY); prope Buitenzorg (Bogor), without date, *Zippelius s.n.* (L); without specific locality and date, *Zippelius s.n.* (NY). MOLUCCAS: Without specific locality and date, *Forsten s.n.* (L); Seram (Ceram), Selamena, ca. 700–800 m, 26 Nov 1917, *Kornassi 559* (L); Boeroe (Buru), ca. 700 m, 21 Feb 1922, *Toxopeus s.n.* (L). SUMATRA: Simalur (Simulue), 14 Oct 1918, *Achmad 667* (K); Dolok Parhorasan, Asahan, in 1936, *Boeea 10548* (NY); vic. Rantau Parapat, Bila, 18–23 May 1932, *Toroes 2217* (NY); Atjeh, Gajolanden, van Gadjah naar Blang Kedjeren, ca. 1450 m, 27 Feb 1937, *van Steenis 9377* (K, NY).

JAPAN. RYUKYU. Okinawa: Yaeyama-gunto, Iriomote Isl., to Inaha from Hoshitate, 7 Sep 1972, *Furuse 1047* (K); Nago-shi, Nago-koen, 4 Sep 1909, *Makino s.n.* (MAK); Amamio-shima, Yamato-mura, Yamatohama, 4 Sep 1910, *Makino s.n.* (MAK); Naha-shi, Syuri, in 1905, *Moriyasu s.n.* (MAK); Yaeyama-gunto, Iriomote Island, along the Urauchi River, 21 Aug 1951, *Walker & Tawada 6681* (L).

MALAYSIA. Kelantan, Kota Bharu, Feb 1917, *Ridley s.n.* (K); Selangor, without date, *Ridley s.n.* (K); Perak, Sungai Kriau Estale, 28 Nov 1937, *Spare 34456* (K); Negri Sembilan, 19 Aug 1916, *Col.*? (K); Penang, Polo Bootong, Jul 1886, *Col.*? (K).

PHILIPPINES. Without specific locality and date, *Cuming 2281* (K, L, MO); without specific locality and date, *Cuming 2461* (L). LUZON: Prov. Sorsogon, without date, *Cuming 484* (K, L); Prov. Tayabas (Quezon), Luoban, May 1907, *Elmer 7729* (MO); Prov. Laguna, Los Banos, Mt. Maquiling, Jun–Jul 1917, *Elmer 18313* (K, L, MO, NY).

SINGAPORE. Botanical Garden, Jan 1925, *Nus s.n.* (BH); Botanical Garden, 5 Oct 1927, *Nus s.n.* (BH, K, NY); without specific locality, in 1830, *Wallich s.n.*(K).

SRI LANKA. Ceylon, without date, *Wallich s.n.* (K).

TAIWAN. Without specific locality and date, *Faurie s.n.* (P); without specific locality and date, *Henry 2027* (NY). HUALIEN: "Hwalien," 16 Oct 1934, *Suzuki 5383* (TAI). PINGTUNG: Pinlang, 8 Oct 1970, *Kuo 173764* (TAI). TAIPEI: Sichih, Oct 1976, *Kuo 6790* (TAI); Taipei, 15 Oct 1935, *Nakamura s.n.* (US); Taipei, 10 Jul 1934, *Odashima 17826* (L, TAI).

VIETNAM. South Vietnam, 1 Nov 1967, *Guong 647* (P). ANNAM: Nhatrang, 17 May 1922, *Poilane 3258* (P). TONKIN: Dich Long?, 9 Oct 1885, *Bon s.n.* (P); "Tonkin occidental," without date, *Bon 6207* (P); without specific locality, Jul 1906, *Moubet 311* (P).

20b. P. dichotomum var. **angustissimum** (Hook. f.) Park, comb. nov.

Polygonum pedunculare var. *angustissimum* Hook. f., Fl. Brit. India **5:** 48. 1886. Type. India. "Polyg. No.

49. Herb. Ind. Or.," *Hook. f. & Thomson* (not seen); "Nilghiris," without date and collector's name (not seen); "The Concan, Belgaum, *Ritchie*" (one specimen believed to be Ritchie's collection from Belgaum at K): Hooker also cited Wight's plate 1802A in Icon. Pl. Ind. Or., 1852.

Polygonum pedunculare var. *nilagiricum* Hook. f., Fl. Brit. India **5:** 48. 1886. Type. India. "Nilgiri Mts. at Conoor," 13 Mar 1870, *Clarke 10810* (two collections (A & B) on the same sheet at K. *Clarke 10810A* is here designated as lectotype, K!; isolectotype, K!).

Polygonum pedunculare var. *robustum* Hook. f., Fl. Brit. India **5:** 48. 1886. Type. India. "Nilghiri and Pulney," collector's name not cited (not seen); Hooker also cited Wight's plate 1802B in Icon. Pl. Ind. Or., 1852.

Stem sparsely to moderately prickly. *Leaves* sessile to subsessile; blades linear lanceolate or lanceolate, rarely narrowly elliptic, 6–12 cm long, 1–2 cm wide, acuminate at the apex, attenuate at the base. *Ocreae* glabrous, weakly to moderately ciliate at the tip. *Inflorescence branches* relatively stout, ascending, densely glandular-pubescent, each terminated by a relatively dense panicle of many flower fascicles; bracts ciliate, moderately to densely hispid. *Achenes* rugose, completely dull.

Distribution. Known only from India (Fig. 35).

This variety is distinct from var. *dichotomum* in having sessile to subsessile, linear lanceolate to lanceolate leaves with somewhat attenuate bases. Ocreae are weakly ciliate at the tip, and achenes are completely rugose and dull. In addition, inflorescence branches are densely glandular-pubescent, and each one is terminated by a rather dense panicle.

Hooker (1886) divided this form into three varieties based on slight differences in leaf shape and size, but no other characters clearly distinguish these three. Since these three names were published simultaneously by Hooker (1886), and since their synonymy has not been previously recognized, I have selected the name var. *angustissimum* (Article 57.2 of ICBN), as it has been more widely used to refer this taxon than the other two.

I failed to locate the type specimen of var. *robustum.* However, from examining Wight's plate cited by Hooker and later collections from the same regions, I can be sure of its identity.

Additional specimens examined. INDIA. Without specific locality and date, *Wight 2460* (K). ANDHRA: Visakhapatnam Dist., Araku Valley, 16 Sep 1961, *Balakrishnan 609* (MAK). MADHYA: Pachmarhi, Little Falls, 14 Jun 1912, *Haines 3488* (K); Pandhurna Valley, 2000 ft, Oct 1918, *Haines 4477* (K); Pachmarhi, Oct 1970, *Johri s.n.* (FLAS). MADRAS: Nilgiris Dist., Sep 1883, *Gamble 12328* (K); Mt. Nilgiri and Kurg, without date, *Thomson s.n.* (L); Nilgiri Mts., without date, *Thomson s.n.* (K); Palni Mts., Sep 1836, *Wight 2461* (K); without specific locality and date, *Wight s.n.* (K, 2 sheets). MYSORE: Jog Falls, Sharavati River above Jog Falls, 14 Nov 1970, *Cook & Gut 197* (K). ORISSA: Koraput Dist., Bambandha nr. Pottangi, 3000 ft, 15 Oct 1950, *Mooney 4186* (K).

21. Polygonum clarkei Park, Brittonia **38:** 217. 1986.

Polygonum pedunculare var. *assamicum* Hook. f., Fl. Brit. India **5:** 48. 1886. Type. India. Jynteapore, 21 Nov 1872, *Clarke 17920A* (lectotype, K!, designated by Park, Brittonia **38:** 217. 1986). Syntype. India. Jynteapore, 21 Nov 1872, *Clarke 17920B* (K!).

Scandent *herb,* 0.3–1 m tall; stem branched, angled, sparsely to moderately armed with recurved prickles along the angles; prickles bristlelike, ca. 1 mm long; lower nodes often producing roots. *Leaves* short-petioled, upper ones usually subsessile; blades linear lanceolate to lanceolate, 4.5–9.8 cm long, 0.8–1.6 cm wide, subobtuse to acuminate at the apex, weakly hastate to subsagittate with small rounded lobes at the base, finely ciliate at the margin; upper surfaces usually bearing appressed bristlelike simple multiseriate hairs toward the apex; lower surfaces glabrous; midveins beneath moderately to densely prickly; petioles very short, 0.2–1 cm long, retrorsely prickly. *Ocreae* scarious, cylindrical, 1.2–2 cm long, oblique, eciliate, often with a few bristlelike hairs along the veins, bearing a ring of prickles at the base, lower ones often lacerate. *Inflorescences* terminal or in the axils of upper leaves, elongate, 5–15 cm long, very slender, branched 1–3 times, often geminate, sparsely glandular-pubescent; glandular hairs stalked, 0.3–0.5 mm long; inflorescence bracts resembling ocreae but much reduced, 3–5 mm long, often weakly ciliate and bearing a few bristlelike hairs along the veins; each inflorescence branch terminated by a somewhat headlike cluster of 4–6 flower fascicles. *Flower fascicles* closely spaced, 2- to 3-flowered; bracts elliptic, ca. 2 mm long, glabrous, ciliate at the tip; bracteoles elliptic, 2–3 mm long, glabrous, ciliate at the tip, fused at the base; pedicels short, 2–3 mm long, glabrous, included in bracteoles; perianth 5-parted, reddish, 2–3 mm long; tepals broadly elliptic, obtuse, not becoming

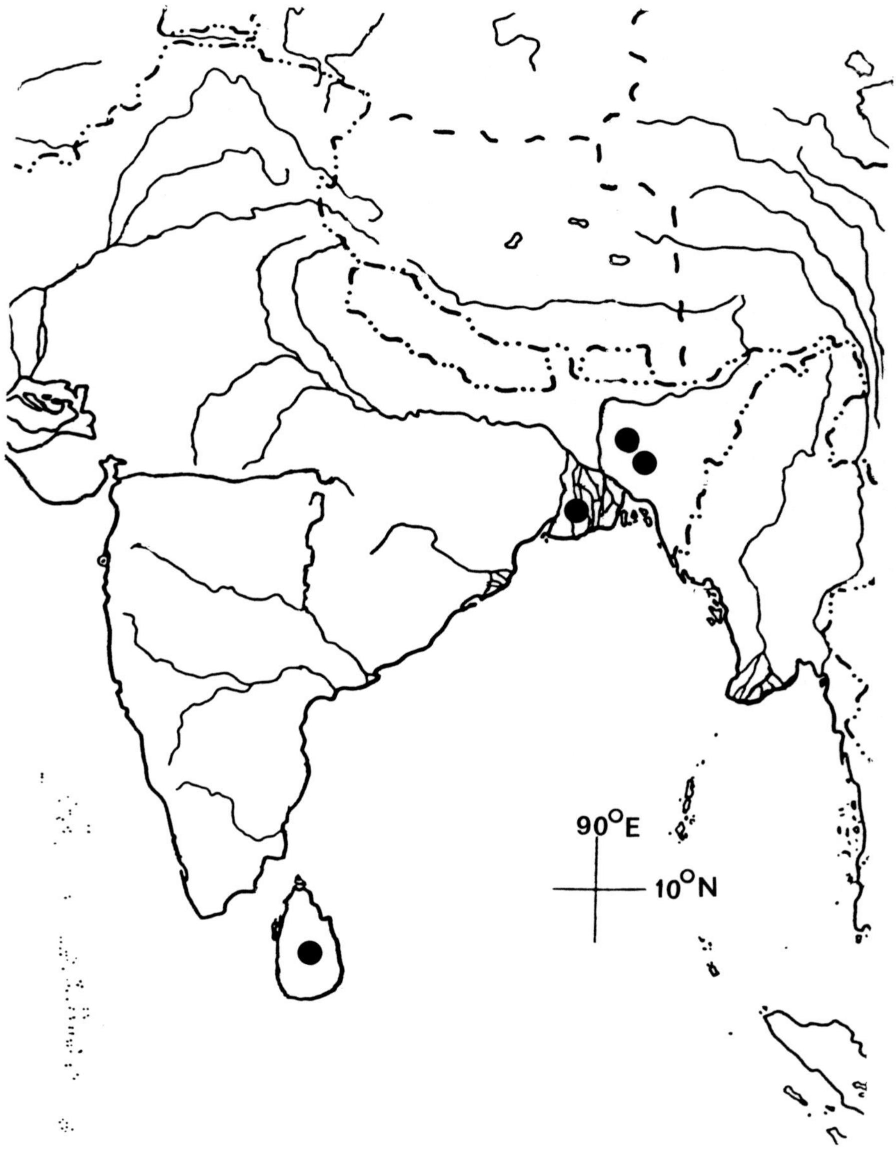

Fig. 36. Distribution of *P. clarkei.*

fleshy; stamens five, inserted at the base of perianth in one whorl, ca. 1 mm long; style one, 3-cleft, ca. 0.5 mm long, included; stigmas three. *Achenes* weakly trigonous, 2–3 mm long, ovoid in outline, acute at the apex, dark brown, rugose, somewhat dull.

Distribution. In northern India, Bangladesh, and Sri Lanka (Fig. 36).

This species was originally described as a variety of *P. pedunculare* (=*P. dichotomum*), but it appears to be distinct from that species in having a three-cleft style and a rugose achene (Park, 1986a).

Additional specimens examined. BANGLADESH. Without specific locality and date, *Hook. f. & Thomson 259* (K); "Plan. Ganget. Inf.," without date, *Thomson & Hook. f. s.n.* (L).

INDIA. ASSAM: Khasi Hills, Burnihat, 200 ft, 1 Jun 1949, *Koelz 22839* (MICH).

SRI LANKA. Ceylon, without date, *Wallich s.n.* (K).

Acknowledgments

This work was mainly derived from a dissertation submitted in partial fulfillment of the requirements of a Ph.D. degree from Cornell University. I would like to express my deep appreciation to the late Dr. M. D. Whalen, my committee chairman, for his continuous support and invaluable suggestions. I also wish to express my sincere gratitude to the other members of my committee, Drs. P. Gregory and P. Marks.

I thank Dr. N. Uhl for her valuable comments on morphological characters, and Dr. D. A. Young for his assistance with cladistic analysis. Drs. W. Dress and R. Barneby shared their vast knowledge in the solution of problems in typification and nomenclature. Dr. P. Hyppio kindly arranged the loans from the major herbaria of the world. I am also grateful to the staff members of herbaria cited in the text, who made generous loans of their collections. For his assistance in editing the manuscript, I would like to express my deep appreciation to Robert Dirig. I am also indebted to Bente King, who prepared some of the illustrations. I would also like to thank Drs. R. S. Mitchell and M. F. Denton for reviewing the manuscript. Finally, I express my sincere gratitude to my family for their understanding, support, and encouragement.

This work was supported financially by a grant-in-aid from Sigma Xi and a grant from the L. H. Bailey Hortorium, Cornell University. Part of this work was completed during my tenure as research associate at the New York Botanical Garden, funded by the Japan–U.S. Cooperative Fund. I extend my thanks to Dr. T. Koyama, the chairman of the Japan–U.S. Cooperative Fund, for his support and understanding.

Literature Cited

Andersen, R. N., W. E. Lueschen & J. R. Zaremba. 1985. Prickly smartweed (*Polygonum bungeanum*), a new weed in North America. Weed Science **33:** 805–806.

Chamisso, L. A. 1833. Polygoneae. Linnaea **8:** 130–140.

Chaney, R. W. 1947. Tertiary centers and migration routes. Ecol. Monogr. **17:** 141–148.

Curtin, P. D. 1969. The Atlantic slave trade. The University of Wisconsin Press, Madison.

Dammer, U. 1892. Polygonaceae. *In* A. Engler & K. Prantl, Die natürlichen Pflanzenfamilien **3(1a):** 1–36.

Danser, B. H. 1926. Die Polygonaceen Niederlaendisch-Ostindiens. Bull. Jard. Bot. Buitenzorg, sér. 3. **8:** 117–261.

Fernald, M. L. 1931. Specific segregations and identities in some floras of eastern North America and the Old World. Rhodora **33:** 25–63.

Graham, A. 1972. Outline of the origin and historical recognition of floristic affinities between Asia and eastern North America. Pages 1–16 *in* A. Graham (ed.), Floristics and paleofloristics of Asia and eastern North America. Elsevier Publ., New York.

Graham, S. A. & C. E. Wood, Jr. 1965. The genera of Polygonaceae in the southeastern United States. J. Arnold Arbor. **46:** 91–121.

Gross, H. 1913. Beiträge zur Kenntnis der Polygonaceen. Bot. Jahrb. Syst. **49:** 234–339.

Hara, H. 1952. Contributions to the study of variations in the Japanese plants closely related to those of Europe or North America. Part I. J. Fac. Sci. Univ. Tokyo, Sect. 3, Bot. **6:** 29–96.

———. 1966. The flora of eastern Himalaya. University of Tokyo Press, Tokyo.

Haraldson, K. 1978. Anatomy and taxonomy in Polygonaceae subfam. Polygonoideae Meisn. emend. Jaretzky. Symb. Bot. Upsal. **22(2):** 1–95.

Hedberg, O. 1946. Pollen morphology in the genus *Polygonum* L. s. lat. and its taxonomical significance. Svensk Bot. Tidskr. **40:** 371–404.

Hooker, J. D. 1886. The flora of British India. Vol. 5. L. Reeve & Co., London.

Kluge, A. G. & J. S. Farris. 1969. Quantitative phyletics and the evolution of anurans. Syst. Zool. **18:** 1–32.

Komarov, V. L. 1936. *Polygonum. In* V. L. Komarov (ed.), Flora SSSR **5:** 594–701.

Kuntze, C. E. O. 1891. Revisio generum plantarum. Pars 2.

Laubengayer, R. A. 1937. Studies in the anatomy and morphology of the polygonaceous flower. Amer. J. Bot. **24:** 329–343.

Léveillé, A. A. H. 1904. Nouveautés Chinoises, Coréennes et Japonaises. Bull. Soc. Bot. France **51:** 422–424.

———. 1912. Decades plantarum novarum. XCIII–C. Repert. Spec. Nov. Regni Veg. **11:** 295–307.

——— **& E. Vaniot.** 1902. Énumération des plantes du Kouy-tchéou. Genre *Polygonum.* Bull. Acad. Int. Géogr. Bot. **11:** 338–344.

Li, H. 1972. Eastern Asia-eastern North America species-pairs in wide-ranging genera. Pages 65–77 *in* A. Graham (ed.), Floristics and paleofloristics of Asia and eastern North America. Elsevier Publ., New York.

Löve, A. & D. Löve. 1956. Chromosomes and taxonomy of eastern North American *Polygonum.* Canad. J. Bot. **34:** 501–521.

Makino, T. 1903. Observations on the flora of Japan. Bot. Mag. (Tokyo) **17:** 117–121.

———. 1909. Observations on the flora of Japan. Bot. Mag. (Tokyo) **23:** 81–95.

Meisner, C. F. 1826. Monographiae generis Polygoni prodromus. Geneva.

———. 1832. Synopsis Polygonearum. *In* N. Wallich, Plantae asiaticae rariores **3:** 53–65.

———. 1855. Polygonaceae. *In* C. F. P. Martius, Flora brasiliensis **5(1):** 1–60, pls. 1–27.
———. 1856. Polygonaceae. *In* Alph. de Candolle, Prodromus systematis naturalis regni vegetabilis **14:** 1–186, 693–695.
———. 1865. Polygonaceae. Ann. Mus. Bot. Lugduno-Batavum **2:** 55–65.
Mitchell, R. S. 1971. Comparative leaf structure of aquatic *Polygonum* species. Amer. J. Bot. **58:** 342–360.
Moore, H. E., Jr. 1973. Palms in the tropical forest ecosystems of Africa and South America. Pages 63–88 *in* B. J. Meggers et al. (eds.), Tropical forest ecosystems in Africa and South America: A comparative review. Smithsonian Institution Press, Washington, D.C.
Nakai, T. 1909. Polygonaceae of Japan. Bot. Mag. (Tokyo) **23:** 411–430.
———. 1926. A new classification of Linnaean *Polygonum.* Rigakkai **24:** 289–301.
Ohki, K. 1926. Polygonaceae of the Island Iki. Bot. Mag. (Tokyo) **40:** 48–58.
Ohwi, J. 1965. Flora of Japan (in English). Smithsonian Institution Press, Washington, D.C.
Park, C. 1986a. Nomenclatural changes in *Polygonum* section *Echinocaulon* (Polygonaceae). Brittonia **38:** 217–219.
———. 1986b. Nomenclatural typifications in *Polygonum* section *Echinocaulon* (Polygonaceae). Brittonia **38:** 394–406.
———. 1987. Flavonoid chemistry of *Polygonum* sect. *Echinocaulon*: A systematic survey. Syst. Bot. **12:** 167–179.
Rafinesque, C. S. 1837. Flora telluriana. Part 3. Philadelphia.
Raven, P. H. 1972. Plant species disjunctions: A summary. Ann. Missouri Bot. Gard. **59:** 234–246.
——— **& D. I. Axelrod.** 1974. Angiosperm biogeography and past continental movements. Ann. Missouri Bot. Gard. **61:** 539–673.
Ridley, H. N. 1930. The dispersal of plants throughout the world. L. Reeve & Co., Kent.
Roberty, G. & S. Vautier. 1964. Les genres de Polygonacées. Boissiera **10:** 7–128.
Small, J. K. 1892. A preliminary list of American species of *Polygonum.* Bull. Torrey Bot. Club **19:** 351–370.
———. 1895. A monograph of the North American species of the genus *Polygonum.* Mem. Dept. Bot. Columbia Coll. **1:** 1–183.
Smith, W. W., S. C. Banerji & M. S. Ramaswami. 1913. Two decades of new Indo-Burmese species. Records Bot. Survey India **6:** 29–40.
Steward, A. N. 1930. The Polygoneae of eastern Asia. Contr. Gray Herb. **5(88):** 1–129.
Swofford, D. L. 1985. Phylogenetic analysis using parsimony (PAUP). Version 2-4. Illinois Natural History Survey, Champaign.
Thorne, R. F. 1973. Floristic relationships between tropical Africa and tropical America. Pages 27–47 *in* B. J. Meggers et al. (eds.), Tropical forest ecosystems in Africa and South America: A comparative review. Smithsonian Institution Press, Washington, D.C.

Numerical List of Taxa

1. *P. perfoliatum* L.
2. *P. senticosum* (Meisner) Franchet & Savat.
   2a. var. *senticosum*
   2b. var. *sagittifolium* (Léveillé & Vaniot) Park
3. *P. debile* Meisner
4. *P. thunbergii* Siebold & Zucc.
   4a. var. *thunbergii*
   4b. var. *maackianum* (Regel) Maxim. ex Franchet & Savat.
5. *P. biconvexum* Hayata
6. *P. arifolium* L.
7. *P. stelligerum* Cham.
8. *P. muricatum* Meisner
9. *P. hastatosagittatum* Makino
10. *P. strigosum* R. Br.
11. *P. breviochreatum* Makino
12. *P. dissitiflorum* Hemsley
13. *P. praetermissum* Hook. f.
14. *P. subsagittatum* (De Wildeman) Park
15. *P. bungeanum* Turcz.
16. *P. rubricaule* Cham.
17. *P. sagittatum* L.
18. *P. meisnerianum* Cham. & Schldl.
    18a. var. *meisnerianum*
    18b. var. *beyrichianum* (Cham. & Schldl.) Meisner
19. *P. brachypodum* Baker
20. *P. dichotomum* Blume
    20a. var. *dichotomum*
    20b. var. *angustissimum* (Hook. f.) Park
21. *P. clarkei* Park

Index of Scientific Names

Page numbers in **boldface** indicate primary page references. Page numbers with an asterisk (*) indicate pages with illustrations or maps. Synonyms are in *italics*.

CHONG-WOOK PARK

Chong-wook Park was born 16 May 1953 in Pusan, Korea. His botanical interests came early and stemmed from his father, who was also a plant systematist. He studied botany at Seoul National University, where he received his B.S. (1976) and M.S. (1978). In 1980, he went to Cornell University to study systematics at the L. H. Bailey Hortorium, and the present work on *Polygonum* section *Echinocaulon* is mainly derived from his doctoral dissertation submitted to Cornell University. After receiving his Ph.D. from Cornell University in 1985, he worked as a postdoctoral museum intern and subsequently as a research associate at the New York Botanical Garden. Currently he is a postdoctoral fellow at the Harvard University Herbaria, continuing his research on the genus *Polygonum*.

134
30848